THIS IS IT

A Manager's Guide to Information Technology

THIS IS IT

A Manager's Guide to Information Technology

John Eaton & Jeremy Smithers

London Business School

Philip Allan

First published 1982 by

PHILIP ALLAN PUBLISHERS LTD
MARKET PLACE
DEDDINGTON
OXFORD OX5 4SE

British Library Cataloguing in Publication Data

Eaton, John
   This is IT
   1. Information storage and retrieval systems
   I. Title   II. Smithers, Jeremy
   001.5   Z6999

   ISBN 0-86003-514-X
   ISBN 0-86003-614-6 Pbk

Set by Midas Publishing Services Ltd, Oxford
Printed in Great Britain at the Pitman Press, Bath

To Anna, Henry, Joe and Neil;
it's their future

Contents

Part VI   The Impact on Society

Preface

Information Technology (IT) is a new label for a collection of old ingredients, technologies which up to now have enjoyed disparate natures and histories. The ingredients have familiar labels: computers, telecommunications, electronics, office products and a host of other more specific items. The reasons for their separate natures have up to now been rooted not only in fundamental differences in technology but also in major external pressures, notably regulatory legislation.

In recent years major shifts have occurred in both of these separating influences, resulting in significant changes in the nature and structure of these industries. The major shift in the technological base has come with the extraordinary developments in microelectronics. The 'microchip' has rapidly become the basic component of a group of industries which previously relied on specialised components to preserve their separate identities.

Whilst the advent of microelectronics may have appeared to be a dramatic and overnight process, in fact the microchip has been quietly evolving over the past twenty years in a very predictable manner. In 1965 Dr Gordon Moore, then President of Fairchild Semiconductors, predicted that the density at which microelectronic circuits could be fabricated would roughly double every year into the foreseeable future (twice as many circuits that is would occupy the same physical space).

Since then this has in practice been the case and indeed is likely to continue into the future with very little slackening of pace. The amount of detail contained in the newest integrated circuits is almost beyond human comprehension. By way of an analogy, imagine a street map for the whole of the Greater London area reduced in size so that four complete maps would be no larger than a postage stamp.

The other major shift has been in the government's attitude to regulation. Three particular events characterise this change. In the UK the British Telecommunications Act (July 1981) which split British Telecom off from the Post Office also significantly liberalised their monopoly position. This Act, primarily influenced by the Beesley Report, has had a catalytic effect on British Telecom.

In January 1982 two major anti-trust cases were resolved in the USA. First, AT&T (the parent of the dominant American telephone system supplier, Bell Telephone) negotiated a settlement with the government which allows it to change from being a telephone company into one offering a range of telecommunications products and services (some of which have a distinctly data processing flavour to them). In return it has had to split up its operating companies into independent concerns.

Second, the US Justice Department dropped its long running (13 years) anti-trust suit against IBM, signalling the end of an era in attempting to contain the mighty giant of the computing world. With its interests in Satellite Business Systems and other telecommunications networks, IBM stands poised to extend its position of dominance in computing over even wider fields.

So what is Information Technology? The Department of Industry defines IT as: 'the acquisition, processing, storage and dissemination of vocal, pictorial, textual and numeric information by a microelectronics-based combination of computing and telecommunications'.

Other attempts at definition have been equally unwieldy. The French use the word *telematique* which somehow seems to capture the flavour of the subject better, while in the USA, Harvard's Antony Oettiger constructed the ghastly term 'compunications'. So it seems that IT is something which is better described than defined, and that is what we are attempting to do here.

The motivation to write this book grew out of a need to provide teaching material for our courses at the London Business School. While there are plenty of good texts on specific aspects of computing, telecommunications, management information systems and so on, most of these are aimed at the systems design or data processing professional. Books that aim to cover this subject for students of management either provide too superficial consideration of the technology or else have rather a narrow perspective in their coverage.

We believe that it is important for managers to have a sound conceptual understanding of the components of Information Technology. For too long managers have been at the mercy of their technical experts, particularly in the data processing world. While this has been in large measure their own fault, it also stems from the attitude of the British educational system which esteems the value of arts and pure science over that of applied science. This may not have mattered so much while the application of Information Technology, in the guise of computers, was relatively limited. However, today the impact of IT on all aspects of business and social life is becoming inescapable, and in many instances is becoming the dominant technological influence on our personal and business lives. No longer can we afford to leave our fate in the hands of technical experts, since if we do they will in effect be taking major social, economic and political decisions about our future as individuals, the future of the organisations which employ us, and in some respects the future of Britain as we know it.

One of the major problems in writing a book such as this, is the rate at which the technology is developing. It is difficult enough to keep abreast of the developments in computing, let alone telecommunications, broadcasting, satellites, image processing, etc. Thus in this book we have tried to describe the basic concepts and principles behind each component of IT, since it is unlikely that these will change significantly over the course of the next decade. What does occur daily is the announcement of new products and processes. However, the lead time between a new line of technology becoming feasible and its widespread application is at least two years. Thus we would not expect to see any major omissions from our catalogue of technologies for the next couple of years.

What change even more rapidly than technology are the issues that relate to it. In the UK 1982 has been Information Technology

Year, and since we finished writing this book, there has been a major debate within the government on two specific issues. First, the Electronics Economic Development Committee (EDC) of the National Economic Development Office (NEDO) has published a blueprint for the electronics industry called *Policy for the Electronics Industry* (NEDO, Electronics EDC, 1982), which provides an excellent review of the industry and of the threats and opportunities facing it; but it is much less convincing in its policy proposals, which appear very much the same medicine as before.

Second, major proposals have been made to recable the country (see *Cable Systems*, A Report by the Information Technology Advisory Panel, Cabinet Office, 1982), in order to provide the necessary infrastructure for the information age (well, at least that part of it dependent upon the availability of 30 or more TV channels in every home). This could be done in a variety of ways ranging from the proposal for a chain of local cable TV companies covering perhaps half the population at a cost of £2.5 billion, to British Telecom's proposal to handle the whole project at a rather more leisurely pace.

Given our belief that a proper understanding of the significant elements and characteristics of the technology is a necessary prerequisite to a full awareness and appreciation of the application of Information Technology and of the major issues that it raises, we make no apology for devoting a significant part of this book to providing you with such an understanding.

The book is organised into six parts, each of which deals with a different aspect of Information Technology.

In Part I we examine the historical context of information processing and attempt to categorise its different aspects.

In Part II we lay out the fundamental concepts of computing from the basic operation of the hardware to the nature and problems of software. We conclude this section by discussing different approaches to file processing.

Microelectronics is the theme of Part III where we look at the way the microchip has developed, the economics of production and consider in some detail current and future applications.

In Part IV we look at telecommunications and the development of different types of network made possible by adopting digital techniques. We examine the converging paths of telecommunications, computing and office products.

Office automation, one of the most hotly discussed facets of IT, is the topic of Part V. We investigate what makes up an office, consider technical solutions being promoted by the suppliers and examine how suited these are to users' real needs.

The first five parts deal with specific technical and management aspects of IT. Part VI deals with the wider social and national implications. Specifically, we discuss the debate on privacy and data protection, and the changing nature of work with its implications for the balance of skills, education and training. We conclude by examining the role of government in stimulating economic activity and its responsibility for aggregate employment, and the place of Information Technology as an agent of technical change in the national economy.

We have also produced a comprehensive Glossary of the particular terms and jargon which abound in this industry, as in every other.

In writing this book we have appreciated the encouragement and criticism of our colleagues, students and friends. In particular, we would like to acknowledge the special contributions of John Steffens for his many suggestions; of Marie Wright for decoding our handwriting and processing the words; of Philip Allan for his confidence in us; and, above all, of Milli and Sheila for their patience and support.

John Eaton and Jeremy Smithers
London Business School, July 1982

Part I

INFORMATION SYSTEMS

Information Technology (IT) has developed in response to a demand for information processing. In many ways this requirement is a relatively recent phenomenon, brought about by the industrialisation of society. A far older, much more basic, requirement is the process of exchanging information. In the following pages we look at the exchange process and examine the effects of automation. We see how this gives us a natural progression into an information processing requirement.

1

The Process of Information Exchange

Information Exchange

In looking at information exchange, it is possible to identify three quite different forms of this process.

First, there is *individual communication,* in which information is exchanged on a one-to-one (or one-to-few) basis. This was the earliest and most basic form of communication and, unlike the other forms, it can be most effectively carried out without the use of any technology at all. Technology becomes necessary only when this communication takes place over any kind of distance. While technology enables this process to be carried out on a wider basis, it in no sense enhances the process. Personal communication is best done face-to-face and all the current technologies seem poor substitutes for this direct contact.

Second, there is *broadcasting* of information, which can be represented as a one-to-many process. There is a vast spectrum of information that may be broadcast that includes facts, opinions, propaganda etc. These exchanges may either be initiated by the transmitter or by the receiver. In the 'transmitter-driven' mode, an individual (or homogeneous group) wishes to transfer some information to a number of individuals (hetero-geneous group). The receivers may be self-selecting or the audience may be specifically targeted. The type of information communicated may range from religious or political doctrine to commercial advertising. In the 'receiver-driven' mode, the receiving group will be selecting or demanding the information from a particular source. Examples of this would be any sort of status reporting or forecasting (e.g. weather forecasts, road traffic reports).

Education falls into both these modes. At one end of the spectrum, authorities decide on the benefits of education and prescribe its dissemination; at the other end, individuals wish to learn a particular subject and acquire the necessary information. This 'one-to-many' process is possible without any technology — for example, at a public meeting — but it is fundamentally a process that has been vastly amplified by the use of technology.

The third form of information exchange is that of *accounting* (with a small 'a'). This word has gathered a quite specific connotation in connection with one commercial aspect of this process, but we use it here in the sense of the exchange of information, in order to keep a record of some activity.

Consider for a moment how we as individuals gain knowledge about a particular event. There are several possibilities: we might experience the event directly or indirectly (say on TV), and we may make some judgment about what occurred in both qualitative and quantitative terms; or we might have the event interpreted by an observer or commentator. In these ways we can gain good qualitative information, though it will tend to be very subjective. However, we are liable to have very meagre quantitative information. For example, if you go to a big football match, you know whether it is very crowded or not; but unless you know the capacity of the ground, it would be very difficult accurately to estimate the number of people watching. The commentator would have a much better idea because of previous experience and knowledge of the capacity of the ground, but we would not expect an estimate to be any more accurate than, say, to the nearest 10%. However, the management do need to know exactly the number of customers present, and so they must measure this accurately.

As Lord Kelvin once remarked, without measurement our knowledge of a particular subject is very meagre. Measurement provides some objective information about events that allows us to make comparisons and classifications. So if we are to find out objective information about some activity, process or event, then we must have records. (This is not to say that the records will necessarily be accurate or unbiased.) The initial gathering of information involves observation, measurement and recording. Once these measurements are recorded, then the information must be classified and structured in some fashion. This is a critical stage as the organising process is bound to introduce subjective elements. To organise the information, it is necessary to have some model or view of how it is to be used and the real-world phenomena to which it relates. However, in execution of this process, original data are liable to be irreversibly transformed.

This third form of information exchange, accounting with a small 'a', has been greatly affected by technology. The process of measuring, recording, classifying and structuring information has been totally transformed by the application of IT.

To summarise, we consider that there are three quite different forms of information exchange, namely:

1. Personal communication (one-to-one)
2. Broadcast communication (one-to-many)
3. Accounting (event/activity to records)

Each of these forms of information exchange uses different types of media and technology. We will be identifying the automation trends in each of these forms and demonstrating that the three streams are being brought together into one interrelated whole.

Information Exchange — The Automation Trend

Personal Communication

Obviously the fundamental and most effective way of communicating is by face-to-face conversation. With the introduction of the telephone, it became possible to speak with another person as far distant as the other side of the world. Though the quality of the sound may not be particularly good, it is usually quite adequate to recognise the caller. The device is very simple to use (or in computer parlance it has a 'simple user-interface'), yet behind this simple instrument lies probably the most complex system yet built by mankind. (The telecommunications system is described in some detail later in the book.) It is now possible automatically to connect oneself to any one of over 700 million people. Yet communication via the telephone represents but a tiny fraction of total personal communication.

Later on in this book we will be discussing how the telephone is becoming the gateway into myriad services, which fall into the other categories of information exchange and handling. Slowly the distinction between categories is becoming eroded. For example, using Prestel (British Telecom's public viewdata service, discussed in Chapter 10), it is possible to leave a message for collection by someone else. Thus we have a system set up for broadcast and information retrieval purposes being used for personal communication.

Speculation about the future of the telephone suggests that it may well become a personal device travelling with us, rather than being fixed in a building. This is feasible with current technology. A recent study by PACTEL (part of the PA Consulting Group) predicts that mobile radio telephones will become very widely available within the next ten years. They may well become standard features of new motor cars before that. Indeed, looking further into the future, two NASA scientists estimated that it would be possible to build communication satellites several kilometres in diameter which would be powerful enough for individuals to communicate with it using a wrist transmitter, Captain Kirk style. Unfortunately the microwaves might also burn a hole in your arm!

Getting back down to earth again, it should be borne in mind that the telephone is a relatively recent invention. The first real communications technology was writing. Using simple materials, it was possible for an individual to communicate messages over long distances, albeit very slowly. This was perfectly satisfactory for many purposes and reflected the general pace of life.

For the major part of the time that written messages have existed, they were delivered individually or personally. When the volume reached a certain level, the distribution system started to be automated by the introduction of collection, sorting and delivery systems. Again, the user-interface is very simple — it is sufficient to know the address, fix the correct stamp and put the letter into an appropriate collection box. Behind this simple exterior a complex infrastructure has developed, although this was until recently based on relatively low technology and a lot of people. The only significant piece of technology to affect this process, from an individual's point of view, has been the typewriter. This did nothing to increase the volume of information being transmitted, as most people type slower than they can write, but it did help the clarity of the text.

A more recent innovation has been the ability to send textual information over the telephone network, either as telex or telegrams. It is interesting to note that in doing this, the user-interface has become considerably more complicated, so that we need a specially trained operator as an intermediary. This cannot be an economic proposition in the long term, and it would therefore be reasonable to expect these systems to become simpler to use and more accessible.

People communicate on an individual basis, not only for personal and social reasons, but also for business reasons and it is here that major changes are likely. This is only possible because business information exchanges, in general, have quite specific objectives — to order some stock, or to check the availability of some item. People seem prepared to adopt new forms of communication if it will enhance the process of meeting these objectives; and a side effect of this will be to reduce the social interaction that takes place as a part of individual business communication.

The effect of technology, then, is to improve the 'quantity' of personal communication. It is now possible to send messages or converse over longer distances, quicker and cheaper than ever before. The earliest technologies were of vastly inferior quality to person-to-person communications, but gradually this quality is improving (*viz*, videophones and teleconferring). However, we do not believe that it will ever be as good quality as direct personal communications.

Broadcast Communication

The earliest way of broadcasting information was by voice, with a speaker or performer and an audience. There are many examples of this today: a sermon in church, a theatrical play, a political meeting. However, the great limitation of this process is the limited number of people one's message can reach at any one time. If the information has only a small potential audience, and they are located in one place, then there is no problem; but many items of information are designed for, or required by, far wider audiences.

Neither was handwriting an effective way of broadcasting information. Texts had to be laboriously copied and so were similarly limited in the audience they could reach. The great breakthrough came when Gutenberg invented the printing press. This was the first time that it had been possible to automate any part of the dissemination of information process. The history of this development is well known, but it is still fascinating to reflect on how rapidly new ideas and beliefs swept through Europe in just a few years after its invention.

Today the printing process is reaching its maturity and will undoubtedly start to decline in the next century. Probably its major drawback, apart from depleting the world of timber, is its basic inability to discriminate. The economies of printing are tied very much to volume production and, as we move towards a more pluralistic world, the demand for uniform products is going to decrease. Printing will have to adjust itself to the philosophy of small batch production and to do that it will need to adopt the technologies that threaten to replace it entirely. Nevertheless, printed information today represents the major part of information that is broadcast.

There has been one recent major development in printing technology which has suited the changing requirements perfectly, namely that of reprographics or photocopying. This allows small numbers of copies of any document to be made without skilled help, rapidly and at a modest cost. In many ways the photocopier tolled the death knell of the privacy of information. Any document within an organisation could be easily copied without anyone being aware that this had been done. The effects of this have been very marked in terms of organisational democracy, as it is now virtually impossible to restrict information within an organisation, and often outside it too. However, the future of reprographics is totally dependent on there being printed (or written) documents in the first place. If, for example, printing as a means of recording information is totally superseded by some electronic form of storage, then photocopying would rapidly follow it into obscurity.

The spoken word, which had been eclipsed by the printed broadsheet, made an outstanding comeback with the arrival of radio. Here was an even more efficient way of disseminating information. While the capital cost of building the radio stations was high, the information could be received by a potentially infinite audience at no extra cost to the broadcaster (the investment in a receiver being borne by the listener). Indeed the word 'broadcast' was seized upon by the organisations operating these facilities and this word rapidly became synonymous with, firstly radio, and subsequently television.

In its turn, radio was outshone by the new star, television. But before this happened, visual information could be recorded and copied in the form of, first still, then moving, pictures. While films serve some of the functions of a broadcast medium, they are very much 'receiver-driven'. People have to go to special places to see them. This is not so with television, once access to a set becomes readily available throughout a society. So television has come to dominate the broadcast media. It requires so very little effort to partake of so much, that it now effectively works in a 'transmitter-driven' mode. People's ability to discriminate is severely questioned, and any information broadcast is sure of a large audience (consisting often of those not particularly seeking this information).

It is not the purpose of this book to offer value judgments on the effect of television, but one can see developments which will both expand its capacity and in many ways change its nature. These changes include offering such a wide diversity of information that the consumer is forced to choose (or will want to choose), so that broadcasting may become 'narrow-casting'. The ability to record programmes for subsequent viewing and 'pay TV' are other ways of achieving this diversity. Another major development is the Warner Brothers' Qube system in Ohio, where each set has a keypad which allows responses to be sent back to the broadcaster. This begins to allow the viewer to interact and thus participate in whatever activity is the subject of the broadcast — although currently this electronic plebiscite is used to adjudicate talent shows! Maybe the ultimate development in this medium was portrayed by Ray Bradbury in his science fiction novel *Fahrenheit 451*. This depicted an environment in which a complete wall of a room was a two-way television set, and became the means of communicating almost everything with the outside world.

'Accounting'

All the information exchange media we have discussed up to now have one thing in common. They only act as 'dumb' (i.e. unintelligent, passive) carriers of information, oblivious to its content and incapable of taking any action based on it. When we come to the accounting process, we stand to gain a very great deal from a technology that can record the information in an intelligent manner, that is in a manner that can subsequently be automatically interpreted and acted upon.

The earliest forms of record keeping had no such facility, since they consisted almost entirely of the quill pen and the ledger book. Only around the turn of the twentieth century did we start devising mechanical methods of storing and processing quantitative information. The earliest systems were very crude, usually recording information by the presence or absence of a series of holes in a piece of paper or card. These holes could automatically be detected by a machine that counted, sorted or tabulated them.

The first major use of 'tabulators' was in the US census of 1890, when they were used to handle the overwhelming volume of information needing to be processed. It was soon recognised that mechanical devices were relatively slow and unreliable and that the future lay with electronics. The major breakthrough was the realisation that information in a 'binary' form could be represented electronically as a simple series of switches.

Years before this, it had been realised by Charles Babbage and Lady Lovelace that it was feasible to build a machine capable of manipulating information, not in a predetermined manner, but by instructions which were a part of the information itself. And so the electronic computer was born. In the next chapter we will review the historical development of the computer and its role in organisations. As we shall see, the computer was not a progeny of the accounting process but rather an adopted child. It is only in the very recent past that it has become mature enough to escape from the confined cloisters in which it has thus far been nurtured.

This escape heralded the beginning of a new era, that of Information Technology. No longer is it possible to classify a piece of technology as being solely applicable to one of the three types of information exchange. Instead, the technology developed for accounting purposes is being applied to personal and broadcast communication (e.g. electronic mail, digital TV, editing machines). Similarly, technologies developed for personal or broadcast communication are being used in the accounting process (e.g. the telephone network and videodisk).

This process is known as 'convergence' and it is most visible in the increasing overlap between the telecommunications, computing and office products industries. In Part IV we will discuss this phenomenon in greater detail.

The characteristics of each medium of exchange, the different technologies applied, and the effects of increased automation are summarised in Figures 1.1a and 1.1b.

	Personal	Broadcast	Accounting
3000 BC	Speech Handwriting	Speech	
		Handwriting	Handwriting
1500 AD		Printing	
1750		Newspapers	
1850	Postal system Telegraph Telephone		
1900		Radio Films	
	Telex		Tabulators Accounting machines
1940		Television Reprographics	Computers (electronic filing)

Figure 1.1a *Development of Key Technologies in Information Exchange*

	Personal	Broadcast	Accounting
Effect of Technology	More quantity, less quality	Amplify	Enable
Complexity for User	Simple to transmit and receive	Complex to transmit, simple to receive	Complex
Type of Exchange	'Dumb' carrier	'Dumb' carrier	Intelligent

Figure 1.1b *Characteristics of Technology*

2

The Development of Data Processing

The previous chapter provided an historical overview of society's development of communications and its growing need for a more sophisticated information processing capability. In this chapter we review the major milestones in the application of computer systems in data processing (usually referred to by the abbreviation 'dp'); in particular, we attempt to relate developments in the basic technology to parallel developments in software applications. The present foundations of the information technology industry are heavily dependent upon the history of the dp and communications industries, both in the structure of the industry and its customers' perception of its nature. The present computing industry, though dominated worldwide by one manufacturer, IBM, with over 65% of the world market, is nevertheless an extremely competitive industry. In contrast, almost since its conception, the telecommunications industry has been either state-owned (as in most of Western Europe), or a very closely regulated monopoly supplier (as until very recently in North America). Now this situation is rapidly changing, and one of the major policy debates of the moment is concerned with exactly how far this deregulation process should, or can, be allowed to go.

Although the focus of this book is on information processing, the historical origins of the computer lie in the development of the calculator, particularly in the work of Pascal, and more especially Babbage and his collaborator Ada, Lady Lovelace. It was Babbage who, though he was never able to realise it in practice, provided in the design of his Analytic Engine, the basic conceptual design for the components of an information processor — input and output units (punched cards and printers), store and processor — and indicated that the instructions to the processor could be changed, i.e. it was programmable. It was Lady Lovelace who, perhaps even more importantly, enunciated the basic concepts of programming — that the machine, however sophisticated, could only blindly follow the program of instructions provided to it.

Electro-mechanical Equipment

However, the first practical data processing (dp) application that would be generally recognised as such today is, in its way, a classic dp application — involving simple, but repetitive, calculations on a vast body of data. The first application was the census returns for the USA, where traditionally, tabulation had been undertaken manually by an army of clerks. The problem was that the clerks had only just finished their work for the 1880 census when the 1890 census was sent out. Herman Hollerith, a statistician employed on the census, devised a machine for automating this process. Like Babbage he used the punched card (an adaptation of the cards used to control Jaquard looms) as his data input medium, but his big step forward was to use an electro-mechanical device, a *tabulator*, to sense electrically the position of the holes punched in each card and then to count and tabulate the data. Hollerith's machines enabled the census to be completed in a third of the time that would have elapsed if the purely manual methods had been used.

From the 1920s through to the 1950s, equipment based on Hollerith's original ideas formed the basis of commercial dp. Data was recorded, stored and processed on punched cards. The processing in the tabulator was fixed by the way it was wired up — it could only be reprogrammed by rewiring, and relatively sophisticated schemes were devised to enable such rewiring to be accomplished speedily. The tabulator was joined by a *sorter* (to sort cards into a specific sequence) and a *collator* (to merge two different sets of sorted cards), enabling basic commercial operations requiring the maintenance of data files to be accomplished. Such equipment was generally known as 'unit-record equipment', in that each unit to be processed (typically a punched card) usually contained one data record.

The Electronic Computer

The Second World War saw a fundamental change in the motivation for computer development. The US government realised the strategic military importance of computers and funded a massive R & D programme. Without this considerable, continuous funding from the US government and associated agencies (e.g. NASA) over the past 40 years, the computing industry today would not exist in its present form. The event which sparked off the original funding was the development of the electronic computer, with its ability for unheard-of speed in calculation, vastly widening the scope of feasible computational problems. Another major attribute of the new machines was that not only did they work at fantastic speed, but they also didn't get bored, lazy or tired! In America, the motivation was the need rapidly to complete the vast number of complex calculations necessary in the development of the atomic bomb and to complete them accurately. In Britain, the motivation was provided by the need of the code-breakers at Bletchley Park to tackle the job of decoding German messages. The needs of nuclear physics, rocket design and space missions meant that this funding was both to continue and grow. From the beginnings in the late 1940s the role of computers has broadened dramatically. This is illustrated in Figure 2.1.

Thus, the initial role was in what we have termed 'problem-solving' — the traditional calculation that the computer was originally designed for. This area has continued to be a major field for the application of computers and is likely always to be so. One of the more interesting developments of the 1970s is the recognition that not only is there a market for microcomputers, but also one for giant machines at the opposite end of the spectrum involved in enormous number-crunching exercises. Soon it was realised that computers could play a major role in processing the volumes of data involved in running a business, and the commercial *data processing* (dp) industry was born. Together with the more recent developments in 'distributed'

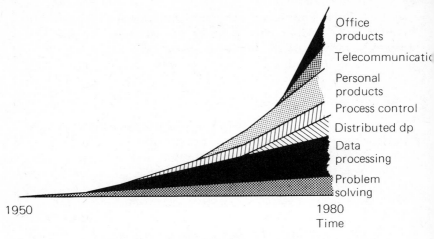

Figure 2.1 *Developments in the Application of Computers*

data processing, this role has turned out to be perhaps the most significant for the computer. But computers have also found roles in many other areas: instrumentation and process control; personal products, such as calculators and watches; in telecommunications; and most recently in a range of office products, especially typewriters and copiers. Traditionally, the computing industry has been seen to be largely concerned with commercial data processing. But now the roles that computers play in processing and storing other types of information are becoming of equal importance, and provide the major focus for this book. However, the historically dominant and significant role of the computer in data processing cannot be overlooked, and provides a natural starting point.

The first electronic computer, ENIAC (which stands for Electronic Numerical Integrator and Calculator, thus beginning the computer industry's obsession with acronyms), ran successfully in 1946, employing more than 18,000 valves, each of which had to function perfectly in order for the machine itself to work. ENIAC was still programmed by direct wiring, and it was later in 1946 that John von Neumann wrote a famous paper detailing the design concepts of a 'stored-program computer' and how it would work, which has been the basis for almost all subsequent computer designs. The first operational stored-program computer was the Mark 1 from Manchester University (in 1948), closely followed by several in America. In America the designers and builders of ENIAC, Presper Eckert and John Mauchly, had left Pennsylvania University and formed the first company to manufacture electronic computers commercially, UNIVAC. The first UNIVAC machine was sold in 1951, but the first UNIVAC sale for commercial data processing purposes was not made until 1954, when a machine was sold to the (American) General Electric Company.

The First Commercial DP Application

In fact, the first computer to be used in business for commercial purposes had been developed in the UK by the J. Lyons Company. Those readers beyond a certain age and with a good memory will remember that the Lyons Company operated a large chain of tea houses. The nature of this business involved a very large number of low value transactions with relatively small margins, so that the company had long been interested in more efficient administration and office management. The senior management of the company became aware of electronic computers whilst on a visit to America, and on their return contacted the group working at Cambridge University under Maurice Wilkes, with a view to using his machine (known as EDSAC, Electronic Delay Storage Automatic Calculator) for business purposes.

One of the major problems they encountered was that the input/output devices used on these early computers were incapable of handling large volumes of data. As we have seen, they had been developed and perceived as very powerful calculators in which, although the calculations might be extremely complex, the volumes of data were very small (business data processing at this time was still using electro-mechanical machines derived from Hollerith's tabulator). However, the problems were overcome and in January 1954 the first Lyons payroll was processed on their computer — the first business data processing application on an electronic computer.

Significantly, Lyons computer was known as LEO — Lyons Electronic Office — and it was certainly perceived by the management of the company at that time as an aid to automation of the office. It is important to appreciate that most of the traditional dp applications (e.g. payroll, accounting, order processing) were originally major office functions — they were composed of structured, formalised, well-understood procedures relatively easy to automate. Present discussion of office automation is directed at functions which are relatively unstructured and not well-understood — and thus more difficult to automate. Ever since this first application in 1954, dp has slowly been automating more and more office operations. As we shall discuss later, present office automation plans are significant because they address those unstructured functions which are presently the prerogative of management, rather than the clerical operations automated historically by dp.

The second half of the 1950s saw a rush of companies following UNIVAC in America and LEO in the UK into the business data processing market. These companies came from a range of backgrounds: the companies manufacturing tabulators, sorters etc., entered the market — IBM in America and ICT in the UK. Other companies came from a background in electro-mechanical and electrical equipment, such as NCR from cash registers, Honeywell and Ferranti from switchgear and control equipment, and RCA, General Electric and English Electric. At this time the main markets for computers were still with the government (and the military in particular), and universities and research establishments. However, the more farsighted could see the vast potential in business data processing.

Stages in the Development of DP

Historically, the development of commercial data processing can be divided into three distinct phases. Although the precise title given to each phase (first, second, third generation) is usually thought of as reflecting purely technological developments, in practice there were other changes, particularly in relation to the role of computers in the organisation, that were occurring in parallel. These phases and their major characteristics are set out in Figure 2.2, together with a fourth column to indicate some of the likely trends in the future. Although the specific dates chosen to represent the boundary of each phase are arbitrary, they adequately indicate the time periods concerned. Many of the terms used in Figure 2.2 are a specialist part of computer jargon — Part II gives an understanding of their meaning.

	1950–60	1960–70	1970–80	1980–?
1. Component Technology	Valves	Transistors	ICs	VLSI
2. System Characteristic	One-off	Mainframe	Mini	Micro
3. Entry Cost (1980 prices)	£4m	£2m	£0.2m	£0.002m
4. Environment	Specialised building	Specialised rooms	Power and air conditioning	Normal office
5. Operating System	None	Batch	Time-sharing	Virtual
6. Application Software	Machine code	High-level language	Packages	Application languages
7. Computer People	Genius	Technical specialist	Application specialist	End-user
8. Application Objective	Problem-solving	Efficiency	Effectiveness	Flexibility
9. Characteristic Applications	Computational problem	Accounts/administration	Main-stream operations	All operations
10. Processing Cycle	Month	Week	Day	On demand
11. End-user Contact	Once removed	Twice removed	Once removed	Direct
12. DP Organisation	Local	Centralised	Distributed	Best-fit
13. DP Budget	Small	Growing	Exploding	Controlled

Figure 2.2 *Historical Development of DP*

Stage 1: The Early Days

In the first of these stages, computers were largely the preserve of the universities and research laboratories, computing 'engines' to drive research. Although the equipment had many undesirable characteristics, the relationship between the machine and the people using it was often good. Specialists within these organisations built and cosseted the machines, and knew how to get what they wanted from them. Typically there was a direct contact between the user (in this case, the researcher) and the machine, with the user having a good technical knowledge of the machine and its capabilities. However, as computers moved into commercial data processing (into the adventurous organisations during the first of our stages, but into almost every large organisation in the industrial world in the second) this close relationship broke down, for several reasons.

Stage 2: The Mainframe Era

The characteristic equipment of this period was known as a *mainframe* computer — it was built on a large central frame — a delicate piece of electronic and electro-mechanical equipment, which had to be housed in a very carefully controlled environment. The equipment (hardware) was physically large, heavy and required as much electricity as a large village in order to function. Almost all of this electricity was turned into heat, which then had to be dissipated. The cost of the equipment alone was colossal — in 1980 prices the cheapest would be about £2 million (by 1970 this entry-level cost would have dropped to about £0.5 million). Given this high investment cost, organisations acquiring this new technology had to try and maximise its utilisation; in practice this meant looking for *high-volume* activities, and *centralised* processing (few organisations could afford more than one of the beasts).

These early applications all tended to be in the accounting and administrative areas, for two main reasons. First, these procedures were well understood (or thought to be) and formalised (accountancy practice imposes a fairly formal set of procedures). Second, in most organisations there were a lot of them (financial transactions), and they were often already processed manually in batches in weekly or monthly cycles. The computer equipment of this era was far from reliable, and the earliest machines seldom worked for more than a few hours before breaking down; once (broken) down, they could stay that way for several days. Consequently, the technology was unsuited to applications requiring processing in 'real time' (i.e. as events occurred). Further, the computers of the day were typically organised to process only one task at a time. Thus the only efficient way to organise this was to collect a large volume of similar transactions (such as time sheets) and process them as one large batch. Not surprisingly, this form of computer organisation was termed *batch processing* and some variant of this is still the most common form of computer organisation.

As they could only afford one computer, and in order to use it efficiently, most organisations began to centralise functions which previously had been dispersed. However, this also gave birth to the rift between the users of the computer's services and the people who ran and looked after it. This situation was made worse by the personnel requirements of the new technology. It was very complex and needed to be supported by highly-trained specialists of a type not previously employed by most organisations. Working times and practices were also different as the computer was often operational 24 hours a day, so that patterns of behaviour in the dp department often became very different from those in the rest of the organisation. Typical amongst these were working strange hours, erratic time-keeping, casual (if not weird) appearance, informality, long hair and large salaries! On top of this, computer experts delighted in talking a language of their own, often using normal English words in a different context. All of this reinforced the sense of alienation between the users and providers of computing services. The dp department often became a business within a business, with loyalties to the computing industry, and motivations apart from those of the rest of the organisation.

During this second stage, users could be, and often were, given an appallingly bad service — viewed as a nuisance to be tolerated, occasionally humoured, but more often ignored. A major misconception by the dp department (which unfortunately is still true in many organisations), was that the world stood still whilst an application was being developed. Many misunderstandings arose. Users would attempt to specify a new application, which the dp department, perhaps after a considerable lapse of time, would suddenly announce had been implemented. The hard-pressed experts naturally became disaffected when told at this stage that circumstances had changed, the world had moved on. Such sequences of events tended to confirm the dp department in its belief that the user could never make up his mind what he wanted, rather than recognise that a changing world was a natural part of the business environment.

In most organisations during this second phase, the dp departments were firmly in control of data processing policy and development. The whole process appeared so complex that senior management typically abdicated all responsibility and the individual user did not have the knowledge or resources to do anything about it. However, the nature of the technology was slowly changing, and during the third of our stages moved the balance back towards the user.

Stage 3: Enter the Mini

This began at the end of the 1960s with the arrival of what are termed *minicomputers* (this was at a time when minis of all sorts were in fashion). These machines took advantage of developments in technology; they were smaller, much more tolerant of their operating environment, and above all were cheaper. On their initial introduction they did not directly compete with the mainframe computer manufacturers and their commercial dp market, as they were designed for laboratory and engineering applications. Originally they had been designed as 'real-time' machines — processing data rapidly, on demand, and usually communicating directly with several 'users' (originally other pieces of laboratory equipment). In a commercial context, this was translated into computer systems which were 'interactive' (they communicated directly with a user at a terminal) and which could service several such users apparently simultaneously.

The traditional mainframe manufacturers tended to dismiss such systems as an irrelevance in the commercial dp world — but they were proved to be wrong and the minicomputer manufacturers, notably DEC (Digital Equipment Corp), HP (Hewlett Packard) and DG (Data General) experienced a decade of exceptional growth. Sophisticated organisations spotted that this was a way round their major problem of an over-centralised dp facility. Minicomputers could be installed as satellite stations, or dedicated to a single application — beginning the trend towards 'distributed' data processing which we discuss more fully later in the book. Interactive computing was also attractive to users, giving them the opportunity to develop applications which required such interaction and for which the processing cycle had to be in hours or less, rather than days. The nature of dp applications began to change from an orientation towards efficiency to one of effectiveness. Computer time-sharing bureaus appeared, giving the user an alternative source of computing resources.

This was the time when the users began to get organised. As the processing cycle of applications became shorter and shorter, so the importance of that processing to the functioning of the business tended to increase (the classic example being the development of airline seat-reservation systems). As a consequence, senior management began to become aware of the problem and set up organisational procedures to attempt to bridge the gap between the users and the dp department. This often took the form of a steering committee, on which the users were a majority, to set dp policy, agree on priorities, and allocate resources; and a project team for each application in which the user had a dominant role. By the end of the 1970s structures such as these were beginning to show success in overcoming alienation.

Stage 4: The Micro Comes of Age?

Now we are entering the next stage in the development of dp in organisations — the arrival of the *microcomputer*. A discussion of the technology that makes this possible, and of its likely impact, constitutes a significant part of this book. Why is the micro-computer so significant? It is so primarily because its cheapness will turn what has been the traditional rationale for dp develop-ment on its head. The equipment is now so cheap that, for some applications, we can virtually ignore its cost; for example, systems may now be justified on five minutes' use per day. Computer systems may become like calculators — always avail-able, but usually with very low utilisation rates.

On the other hand, creating software (the instructions telling the computer what to do) is becoming prohibitively expensive, both in terms of maintenance of existing systems and develop-ment of new ones. Much software development has been aimed at improving the utilisation rates of the hardware, which is now becoming less and less important. It is interesting to note that the same derision and scepticism is shown to today's micro-computers by some dp professionals as was shown to mini-computers and time-sharing systems at the end of the 1960s. This is partly because the professionals feel threatened by a development that may remove some of the rationale for their existence. They welcome radical change no more than most other members of an organisation.

One of the positive aspects of the microcomputer is that it dramatically reduces the entry-level costs — almost anyone can now become a player in the game. Getting into the computing profession used to have all the subtlety of 'Catch 22': if you had experience, then everyone would hire you; if you didn't, no-one would. As a result, the computing profession is narrowly based in terms of the background of its members. But now the availability of 'micros' in schools is demonstrating that computing skills can be acquired at a very early age, and do not necessarily have any correlation with more traditional measures of academic ability.

The great advantage of the microcomputer is its potential to be *personal* — it can be used successfully for a single application or by a single individual, making it a very flexible sort of technology from the organisation's point of view. It potentially cuts out many of the intermediaries in the dp department, but for many applications a division of labour is still sensible. But the dp department will tend to feel threatened by its loss of monopoly knowledge about computer systems. However, the productivity gains which appear to be possible by using this technology will tend to mean that the user comes into direct contact with the computer system, so that control of the technology will increasingly be taken back by user management.

A big question posed by this sort of scenario is how to develop the necessary software in a cost-effective manner — both in terms of development and ongoing costs. There are two broad schools of thought, which are not mutually exclusive. One suggests that the answer lies in the widespread use of packages — we can no longer afford tailormade software and must adapt to what is available 'off the shelf'. Only in this way can the enormous development costs be spread over enough users to give an affordable unit cost. The other believes that packages are too rigid, and that the answer lies in providing more high-level tools — such as application languages (e.g. those already available to financial planners, or design engineers), or more general aids such as English-like query languages for 'accessing' (reaching into) databases. In practice, some combination of both will be appropriate, depending upon the applications within the organisation. We discuss some of these issues in greater detail, both in Part II, where we discuss software in general, and in Part V in the particular context of office automation.

Computers and Organisations

During the second two stages of their development, three characteristics of computers and the data processing industry arose, which subsequently have caused it and its customers immense harm, and which only now are beginning to be dispelled. First, there is the image fostered by both the industry and the media of the computer as a miraculous, superhuman machine, capable of almost anything, but very intimidating and mysterious, to be understood only by those with a PhD in Mathematics. Whilst the computer is indeed a very sophisticated and complex machine, an understanding of the basic concepts of a computer system is nevertheless not beyond the understanding of that famous character 'the average intelligent layman'. For example, many people have a good conceptual, if not detailed, knowledge of how, say, a car or aeroplane functions and of the role they play in transportation systems; yet in their own way they are no more complex than a computer. Yet this lack of a basic understanding, in truth perhaps more a major misunderstanding of the components and capabilities of a computer system, is now a significant social, educational and management problem. Too many organisations have relied upon dp 'experts' to make technological judgments and decisions which have major *business* impacts. The user managers concerned provide little or no inputs, and little attention is given to the business as opposed to technical dp aspects of the decision.

Second, stemming in part from this misunderstanding of its true capabilities, and in part from the commercial necessities involved in developing a vast new market, the computing industry grossly oversold itself, its products and its capabilities. Particularly during the 1960s, tales of the excesses of computer salesmen were legion — the miraculous computer would be a panacea to solve all the problems of the business, and the customers all too often believed it! The result is that the computer industry as a whole probably has a worse public image than almost any other. There still persists a considerable degree of scepticism about the claims of the computer industry in general and their own dp departments in particular, amongst the senior managers of many organisations. Unfortunately, there is some evidence that the painful and expensive lessons learned by the large companies during the 60s and 70s are not being passed on to the smaller organisations presently becoming first-time computer users. One still hears far too much emphasis on the physical hardware aspects of an application, rather than the software and, perhaps most important, the impact on the organisation and the people in it.

Third, the introduction of a miraculous, mysterious, apparently very complex and expensive machine, requiring a considerable amount of cossetting in its operational environment and what might be termed as tender loving care from those looking after it, naturally led almost all of the organisations acquiring computers for general data processing to set up a centralised dp department. Given management's perception of the machine (and usual lack of any real understanding of its capabilities), and its cost, such a response was understandable. Thus, whatever the general organisational structure adopted, dp was almost invariably a centralised facility — often physically remote from the user departments, certainly keeping its distance from individual users, staffed by specialists who perceived their loyalties and career paths to be within the dp industry as a whole, rather than the organisation currently employing them. This attitude was inadvertently fostered by most organisations who kept dp management separate from the rest of the organisation and provided few, if any, routes for dp managers to move into the general management mainstream.

Whilst this degree of sophistication might have been an organisational cost that could have been borne at the time, as we have progressed through the 70s and move into the 80s it is perceived as being increasingly inappropriate and in some instances a major constraint on organisational development. Throughout the past decade the tendency has been towards a greater and greater degree of decentralisation of decision making in organisations, making managers as far as possible responsible and accountable for the attainments of their goals and objectives. A centrally organised and controlled dp operation has become an increasing source of frustration in such circumstances. This has been compounded by the developments of cheaper mini and microcomputer systems; and the movement towards on-line, interactive computing, which brings the end-user into much closer contact with the computer system, and removes many of the buffers and barriers built up around the dp operation. In most organisations these phenomena appear as a movement towards some type of distributed data processing (usually referred to as 'ddp') — typically an attempt to make the organisation of dp more closely mirror the decision-making structure of the organisation as a whole. However, there are many problems and pitfalls in moving from the concept of ddp to the reality — both technical and organisational — which we will review in more detail in Chapter 9.

First, though, in the next part of this book we will attempt to provide you with an appreciation of the basic components of a computer system and their function, as well as an understanding of the meaning of the jargon and 'buzz-words' shrouding computers.

Part II

COMPUTER FUNDAMENTALS

We aim here to provide you with a basic understanding of how a computer works. Although a significant theme throughout this book is to identify and emphasise the software component of systems and, above all, the role of the people who have to use them, this doesn't mean that the technology is irrelevant or that its significant characteristics can be overlooked. In our view, it is of critical importance for user management to have a basic understanding and appreciation of the capabilities and characteristics of the relevant technologies. Only on this basis can informed, intelligent decisions about the potential business impact of such technologies be made. It is perhaps appropriate to reiterate that the primary focus of this book is to provide an insight and understanding of the developments in information technologies, their impact on organisations, their business and their employees at all levels, and on the socio-economic environment in which they live and operate. A proper understanding of the technological base is critical to this.

3

Hardware – The Physical Equipment

What is a Computer?

At the heart of every computer system lies a machine which is able to obey our instructions (telling it how to manipulate some data) at a truly phenomenal speed — relatively commonplace minicomputers may well be capable of obeying more than 1 million instructions per second, whilst the industry giants can process at rates of about 100 million instructions per second. For example, it takes about one-tenth of a second for you to blink your eye — in this time a minicomputer might have executed 100,000 instructions! Assuming we can feed it with instructions and the necessary data at a fast enough rate, and similarly accept the results of its processing, then this machine will happily carry on processing say, 1 million instructions each second, 24 hours a day, 7 days a week, without getting tired or bored, making mistakes, or suffering from any other of the ailments that eventually affect humans engaged in continuous work. We will discuss the form and nature of the instructions that the machine needs in the next chapter — in this chapter we describe the basic physical equipment, the *hardware*, in a computer system. The machine that we have just described, the *processor*, is only a part of our computer system — we need to have a lot more equipment linked to the processor in order to try and satisfy its voracious appetite for ever more instructions and data.

For example, we need equipment with which to enter our instructions and data and to display the results of the computation — known as *input and output devices.* The processor itself requires an associated storage area in which the instructions being executed and data being processed are stored; this is known as the *primary memory* and, together with the processor and some logic circuits enabling the processor to communicate with other pieces of equipment, is referred to as the *central processing unit (or cpu).* There may be many other pieces of equipment connected to the cpu, apart from the input and output devices already mentioned — for example, almost all computer systems require a large file storage area in which instructions and data not presently being worked on may be stored. As another example, it is increasingly common for our computer system to be able to communicate directly with other systems over telephone lines; this too would require a further item of equipment. These other items of equipment around the cpu are known as *peripherals.* A general diagram of a computer system is shown in Figure 3.1.

We have already made use of the terms computer and computer system — is there any significance in the use of one or other of these terms? In everyday use the two terms are probably used synonymously; however, conceptually there is an extremely important distinction which it is worthwhile elaborating upon for a moment. Strictly speaking, a computer is a calculator, a machine which can perform many calculations automatically. It is what lies at the heart of a computer system: the central processing unit, which is a physical piece of equipment.

A computer system is a much more significant concept — it includes all the necessary physical equipment (hardware) and instructions (software) to enable our machine to perform useful, user-oriented tasks. If one looks at the analogy of a motor car, then the processor (computer) would be the engine without which the car would be useless; but the whole car would be equivalent to the computer system and in fact the engine might be a relatively small component. A computer as a physical piece of equipment is in general of little direct use to the end-user — a computer system is defined from the user's point of view, as whatever is necessary to be able to handle his problem.

In this chapter we will be discussing the physical equipment

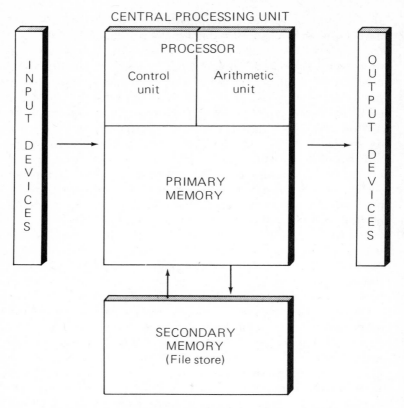

Figure 3.1 *A Schematic Diagram of a Computer System*

that goes to make up a computer system. However, as many of you may already be aware, it is the instructions that the hardware obeys that tend to define the usefulness of the system to an end-user. The end-users are primarily interested in what way the system can support them in performing their jobs — and this is typically directly related to the software capabilities, with the hardware setting some absolute capacity and perform-ance limits. We discuss software, and its interaction with the hardware to form a computer system, in the next chapter.

Input/Output Devices

Thus two very important parts of any computer system are what are known as input and output devices — pieces of equipment enabling us humans to communicate with the processor. They allow us to enter our instructions and data, and to receive the results of its processing. Such machines have to transcribe information in human-readable form — that is, in a form intelligible to humans, such as data on a stock sheet — into a form that a machine can interpret. For example, we might punch holes in cards or paper-tape, using some accepted coding system, to represent both instructions and data. We would then enter these into our processor using a card-reader or a paper-tape reader — a machine capable of sensing where we had punched holes in our cards or tape, and transmitting that information electronically to the processor. Similarly, the processor could transmit the results of its calculations to a printer having the capability to transform the electronic information sent by the processor into human-readable characters, and to print them. Thus, card-readers and paper-tape readers are referred to as input devices, whilst a printer is an output device.

Some pieces of equipment fulfil both an input and output function (hence the usual terminology is to refer to input/output, or *i/o devices*). For example, the increasingly familiar video display unit (vdu) fulfils both roles; the keyboard is an input device, sensing which keys you have pressed and sending that information to the processor, and the display is an output device, accepting electronic information from the processor and displaying it in human-readable form on the screen.

Increasingly i/o devices are becoming the critical part of any computer system. This man-machine interface may prejudice a user's whole view of the computer if it is unsuitable. For example, light pens, graphics tablets, touch screens or voice input may be an easier way than keyboards for people to communicate with machines. Similarly, colour displays with graphical capability or synthesised voice may be a better way for the machine to communicate with people. Figure 3.2 lists the major input/output devices.

The Functions of the Processor

The processor itself is not a monolithic unit, but has three components (as illustrated in Figure 3.1) — an arithmetic unit, a control unit and a storage area — as well as the interfaces

Input	Output
Punched card-reader	Lineprinter
Punched paper-tape reader	Display screen
Keyboard	Plotter
Bar-code reader	Magnetic tape, disk
Optical character reader (ocr)	Voice synthesiser
Magnetic tape, disk	

Figure 3.2 *Input/Output Devices*

necessary to communicate with input/output and other devices. The *arithmetic unit* is exactly as described — a component capable of performing arithmetic operations electronically, and of performing them very rapidly. An analogy could be drawn between the processor and the personal electronic calculator that you probably own. The storage area, known as *primary memory,* is the working memory of the computer system, in which the instructions to be followed, the data to be processed and the results of the processing are stored. Continuing our analogy, it is comparable to paper and pencil on our desk, possibly used to record the form of the calculations, initial data, intermediate results, and the final answer resulting from the use of our electronic calculator.

The *control unit* fetches instructions in sequence from the primary memory, decodes them and acts on them. For example, a sequence of instructions may be to move a number held in a particular location in primary memory to the arithmetic unit, then to move a second number from another location to the unit, then to add the two numbers, then to move the resulting sum to a particular location in primary memory, and so on. In terms of our analogy, this is equivalent to your role in entering data into your calculator in the correct sequence, pressing the keys for the calculations that you require, and noting down intermediate results and the final answer.

It is perhaps not widely appreciated that the capabilities of the arithmetic unit in a computer are extremely limited — basically all it can do is add (or subtract) two numbers, or compare them to find the larger or smaller. Other more complex arithmetic operations have to be built up from these basic capabilities. Let us briefly examine how the hardware actually stores and processes our instructions and data; this involves you in understanding a few straightforward characteristics of the binary number system. It's at this point that the faint hearts give up — but if the majority of children in their early teens can manage it, surely you can! And once you are over this hurdle, it's all downhill. As we emphasised earlier in the book, we are primarily concerned with the impact of information technology — but without a thorough understanding of the technology itself, we feel it to be impossible for you fully to appreciate its impact. And representing information *digitally* — by coded sequences of binary digits — is the key common factor underlying recent developments in information technology.

How the Hardware Works: The Binary System

One of the fundamental factors differentiating an electronic computer from the previous mechanical and electro-mechanical calculators and tabulators is that it can process instructions and data represented only by using the binary number system. In the binary number system there are only two digits, 0 and 1, compared with ten in the decimal system, 0, 1, 2, 3, 4, 5, 6, 7, 8, 9. In the decimal number system with which we are all familiar, each position in a number indicates a power of 10. For example, the number 321 could be represented as follows:

$$1 \text{ times } 10^0 = \quad 1 \text{ (any power of zero} = 1)$$
$$\text{plus: } 2 \text{ times } 10^1 = \quad 20$$
$$\text{plus: } 3 \text{ times } 10^2 = 300$$
$$\overline{}$$
$$\text{total:} \qquad 321$$

Thus in the decimal system, each position has the following value:

power of 10 :	etc . .	10^4	10^3	10^2	10^1	10^0
value :	etc . .	10000	1000	100	10	1

In an exactly analogous fashion, each position in the binary system represents a power of 2, thus:

power of 2	:	etc . .	2^4	2^3	2^2	2^1	2^0
value	:	etc . .	16	8	4	2	1

For example, the decimal number 9 would be represented in binary as 1001, thus:

$$
\begin{array}{rr}
1 \text{ times } 2^0 = & 1 \\
\text{plus: } 0 \text{ times } 2^1 = & 0 \\
\text{plus: } 0 \text{ times } 2^2 = & 0 \\
\text{plus: } 1 \text{ times } 2^3 = & 8 \\
\hline
\text{total: } & 9
\end{array}
$$

You may like to verify for yourself that decimal 321 would be represented in binary as 101000001.

We can perform arithmetic in any number system, binary, decimal, or any other — all we have to remember is at what point a value has to be carried to or from the next highest position. For example, in the decimal system if we were to add together 5 and 7, then we now know automatically that one unit of ten has to be carried to the next higher position to give the answer 12. (Though it is automatic now, as school children we did have to learn this!)

As we have noted, in a binary number system there are only two states, 0 and 1, so that on any occasion in which two binary 1s are added, a carry to the next position will be generated. For example, suppose that we add 5 and 7 together in binary; decimal 5 is equal to binary 101 and decimal 7 to binary 111. The sum is:

$$
\begin{array}{r}
101 \\
+ 111 \\
\hline
1100
\end{array}
$$

i.e.

$$
\begin{array}{rr}
0 \text{ times } 2^0 = & 0 \\
\text{plus: } 0 \text{ times } 2^1 = & 0 \\
\text{plus: } 1 \text{ times } 2^2 = & 4 \\
\text{plus: } 1 \text{ times } 2^3 = & 8 \\
\hline
\text{total: } & 12
\end{array}
$$

Thus we are able to perform simple arithmetic using the binary number system, and are able to specify the logical rules for the performance of that arithmetic.

The importance of the binary system from the point of view of the electronic computer lies in the fact that it has only two digits, or states, which can very easily be represented by a switch on or off; an electrical current flowing or not; the presence or absence of an electrical charge; the presence or absence of a magnetic field. The speed of the electronic computer is derived from the speed at which electrical pulses travel — the speed of light (186,000 miles per second!) — and from electrical switches operating enormously faster than their mechanical equivalents. The significant developments in the technology of processing and storing binary coded information electronically are discussed in Chapter 6.

The movement of electrical pulses at high speed between all of these components requires some very careful synchronisation. Thus a very important component of a cpu is some form of clock! As the electrical pulses pass to a device, the identification of whether a binary one or zero has been received requires the measurement of a time interval, and the determination of whether a significant voltage (compared with some reference value) was detected during this given time interval (thus representing a binary one).

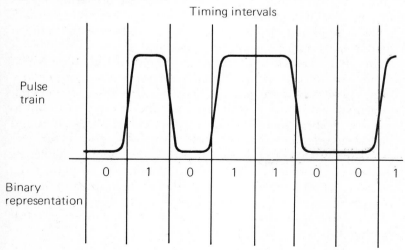

Figure 3.3 *Electrical Representation of Binary Digits*

Schematically this is illustrated in Figure 3.3. For any processor you will see a reference to its *clock rate* — the frequency at which the basic reference pulses will be generated. The higher the clock rate, potentially the faster the processor will operate, though whether this potential power can be tapped for the user will be discussed later.

Thus electronic hardware can store binary data, or instructions coded as binary data, in either an electronic (i.e. the presence or absence of an electrical charge) or magnetic form (i.e. the presence or absence of a magnetic field). Further, it can transfer such binary coded information as a sequence of electrical pulses at a very high rate between different pieces of equipment. And, finally, it can electronically perform elementary arithmetic operations on such binary coded information, again at a very high speed.

File Storage Equipment

To return to a discussion of the hardware components of our computer system, there is one major component we have not yet looked at — the file storage area. As described above, the primary memory of the processor has to hold the instructions and data which the processor is following and operating on; yet further consideration of the analogy of you in your office should make it apparent that there is also a need for other storage media. For example, using the office analogy, the desk surface represents tasks which are currently being worked on (of which there may well be several different items) and other information, typically of a reference nature (telephone directories, diaries etc), which you know full well that you are almost certain to want to refer to at some point during the day. Other information and data is stored in files, each file having some known structure (for example, alphabetic sequence, or date order), which enables its contents to be searched in a logical manner, and for items to be retrieved. Similarly in our computer system, in addition to primary memory there is a requirement for some other type of storage media, capable of holding very large quantities of information; this is known as *secondary memory*.

Magnetic Tape Storage

Secondary memory uses magnetic media to record information in a machine-readable format. The magnetic recording media is available in two different classes of device — magnetic tapes and cartridges, and magnetic disks. On a magnetic tape the information to be stored is recorded in sequence as the tape moves past the read/write head. The operation is exactly analogous to that in a domestic tape recorder (except that the tape itself is usually wider). Information recorded overwrites what was previously stored on the tape; the information on the tape may be read back (e.g. played) many times; and to retrieve any individual item of information, it is necessary to read through all of the physically preceding information on the tape in sequence. Operationally, magnetic tape has the characteristics of cheapness and robustness, rapid data transfer, but slow average access time to any individual data item. Diagrammatically, the storage of information on magnetic tape can be illustrated as shown in Figure 3.4.

The tape can be thought of as being divided into nine equal strips running down its length (but invisible to the eye) known as *tracks* (hence 9-track tape). Each track can hold one bit (binary digit); thus nine bits can be written in rows across the tape. Clearly, the closer the rows of bits are together, the more information can be recorded on the tape — this is known as the tape recording density and is measured by the number of bits per inch (bpi). The present industry standard is 1600 bpi, though older units operating at 800 bpi are still common, and units recording at 6250 bpi are now appearing on the market. However, you are probably saying: what does this mean in terms of storing data that I can comprehend — names, addresses, part numbers, invoice amounts, and so on? In order to answer this sort of question we will have to look in a little more detail at how the storage areas of our computer system are organised.

Units of Computer Storage

The smallest unit of information stored or processed in a computer is a binary digit (i.e. zero or one), known as a *bit*. However, this is too small a unit of storage to be useful to the

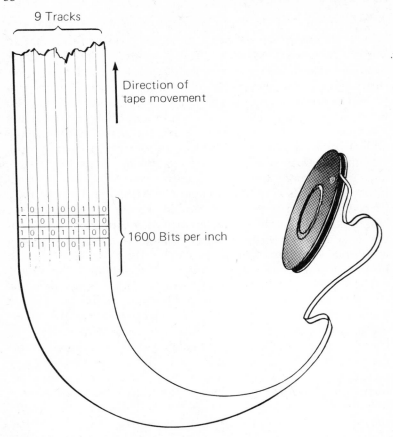

Figure 3.4 *Storage of Data on Magnetic Tape*

normal user. Typically, users wish to store and process characters (actually, groups of characters — names, addresses, part
codes, etc) and numbers (or quantities — invoice amounts, stock
levels, etc). Thus, from a user point of view, the smallest basic
unit of information represented in this way is a character, i.e. the
capital letters A–Z, small a–z, digits 0–9 and special symbols
such as £ $ () + − / and so on. In computer terms, all possible
characters can be represented by a group of 8 bits, known in
computer jargon as a *byte*. For your purposes, a byte and a
character are synonymous.

A byte is essentially a unit of computer storage and transmission; for example, the secondary memory of a computer system is measured in terms of so many thousands (K) or millions (M) of bytes, and the speed at which a terminal transmits input to, and receives output from, the processor is measured in terms of so many characters per second (cps). With 8 bits it is possible to represent the numbers 0–255. This provides more than enough codes for all the numbers, letters (large and small) and special characters. But before this can be done, a recognised coding system must be specified.

There are two common ways of coding the bits in a byte to represent characters: ASCII (American Standard Code for Information Interchange), which is an internationally adopted standard code; and EBCDIC (Extended Binary Coded Decimal Interchange Code), which, because it is the code used by IBM in their computer systems, has become a *de facto* standard code. Generally speaking, you do not have to worry about how a particular instruction or piece of data that you input to the computer is translated into binary; but it is important to be aware that this system function is undertaken for you, and in addition it provides a commonly used standard for information interchange between computer systems.

Magnetic Disk Storage

Magnetic disks can take many forms, but generally comprise one or more flat circular surfaces in a vertical, spaced stack, very similar to a gramophone record, where each surface is coated with a material capable of being magnetised. A read/write head at the end of a movable arm can position itself over any portion of the surface. Logically, this surface is divided into many concentric storage areas, known as tracks, and the read/write head can directly position itself over any one of these tracks. The operations of a disk drive are illustrated in Figure 3.5.

The recording density of a disk drive is measured in two ways, by the number of tracks per inch (typically 20 to 50) and by the number of bits per inch of track (presently about 25000). To you, the capacity is presented as being so many K (thousands), M (millions, or mega) or G (giga, or thousand millions) of bytes, either per drive or in total for your system. Many disks now operate inside units which have been hermetically sealed at the time of manufacture, so that the head can literally 'fly' closer to

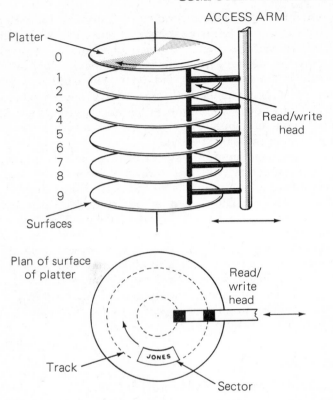

Figure 3.5 *Components of a (Multiple Platter) Magnetic Disk Unit*

the disk surface, thus leaving a narrower magnetic footprint, enabling more tracks and thus more information to be stored in a given area; the first developer of such units was IBM, who gave them the internal code name of 'Winchester'; thus such disk units are commonly referred to as 'Winchester drives'.

Increasing amounts of storage can be placed in smaller and smaller boxes, so that even small entry-level business systems often come equipped with disk drives which have a capacity larger than any drive available 15 years ago, and in about one tenth of the space. Traditional large disks in computer systems are made from rigid 14-inch platters of aluminium — in packs of between 1 and 8 platters per spindle. At present, something of the order of 60 million characters can be stored on one of these platters. With the development of more sophisticated terminals, word processors and personal computers, a new disk storage medium was developed — the flexible or floppy disk (or diskette). In this a 5¼ or 8 inch diameter vinyl disk is spun inside a protective cardboard envelope. The concept and operation of a floppy disk unit are the same as for its larger brothers. On the 5¼ inch floppy, up to 350K characters may be stored; and on the 8 inch disk, up to 1M characters. The latest development is the production of small (5¼ and 8 inch) hard disks, using either one, two or three platters on a spindle, with storage capacity from 10 to 30M characters.

Price/Performance of Storage Media

In terms of both operational performance and cost, there are major differences between primary memory and the two forms of secondary memory commonly found in computer systems. In the primary memory area of the computer, any storage location can be accessed directly in order to retrieve its present contents, or to place new information in it. In the jargon, each location has a unique *address*, which can be used to access it directly. The primary memory area can be thought of as comprising thousands (or millions) of pigeon holes — each of which is big enough to store just one character (byte) — the identifying sequence number of which constitutes its unique 'address' (i.e. it provides a mechanism for uniquely identifying each pigeon hole).

Originally, such memory comprised thousands of minute ferric cores (hence the name 'core' memory originally used to describe this type of memory), each of which could be individually magnetised to represent a binary zero or one. Now such memory is made up using semiconductor components, either *Random Access Memory* (RAM), which would be better named read/write memory; or *Read Only Memory* (ROM), which, as its name suggests, cannot have information written on to it only read from it. ROM is typically used to store very frequently used instructions permanently, which are coded into it at the time of manufacture (these semiconductor devices will be discussed further in Chapter 6).

To a significant degree, the overall speed of a computer is determined by the time it takes the control unit to locate and fetch an instruction or data item from main memory. This is known as the *memory cycle time* and in modern computers is typically measured in micro- or nano-seconds (i.e. millionths, or thousand-millionths of a second). Because it has to operate very quickly, and because each memory location has to be directly addressable, primary memory is relatively much more expensive than secondary memory (typically by a factor of 10 or more). Note that this comparison is in terms of relative cost; as we shall discuss later in Chapter 6, one of the most significant developments in the technology over the past decade is that all memory prices have been falling in absolute terms.

In the case of secondary memory, either tape or disk, information can only be stored or retrieved in blocks, each block containing several items of information. Individual items of information can be accessed by copying a block from secondary memory into primary memory, where each individual location can be addressed. That is, addressing in secondary memory depends upon the way in which that storage area is organised, and in any case can only be carried out for blocks of data. There is an important difference as to how the device physically blocks out the available space (for example, a track on a disk); and the logical blocks that a particular application may employ to structure the way in which its data is stored. We will discuss this point further in relation to file systems and database management in Chapter 5.

Further, as we have noted, primary memory is a purely electronic component incorporating no moving parts; it is a 'solid state' device, whereas secondary memory devices are electro-mechanical. Both tape and disk drives incorporate major mechanical components: in the case of tape, the physical movements of the tape past the read/write head; and in the case of the disk, the rotation of the disk, and the positioning of the arm over a specific track. It is these mechanical operations which result in secondary memory being less reliable and slower in operation than primary memory.

The storage capacity of secondary memory is virtually unlimited, in that as many reels of tape or disk packs may be stored as desired. However, it is very important to distinguish between the drive units on which the tapes or disks are mounted so that the information stored on them can be accessed by the processor, and the tapes and disks themselves. Although the number of tapes and disks may be virtually unlimited, in any particular computer system the actual number of drive units will be very limited; so that at any one point, only a few actual tapes or disks may be directly available ('on-line') to the processor. Thus the secondary storage capacity of a particular computer system is usually specified as its on-line capacity. As we have already mentioned, a discussion of the principles of file organisation on tape or disk and alternative processing methods will be deferred until Chapter 5.

Representing and Processing Numbers

Finally, before moving on to a discussion of the software necessary in order to make all this equipment function usefully for us, there remains one major characteristic of the way computers operate that we must examine. In our previous discussion of the units of computer storage, we illustrated how a byte, representing a character, is the basic unit used. Clearly, for a lot of information-processing applications, a character is a relevant storage and processing unit (for example, for information that would naturally be represented in character form, such as names and addresses, or any textual material).

However, in many instances we wish to store and process numbers as numbers (i.e. not as series of numeric characters), and it is necessary to illustrate how a computer handles numbers. This involves a third unit of storage, namely, a *word*. Basically word and word length are different ways of looking at the same thing — the raw capabilities of a processor. Like a byte, a word is a group of contiguous bits (i.e. a group of adjacent bits, operated on as a whole); but whereas for all practical purposes a byte is always composed of 8 bits, a word differs in size (i.e. its length as measured by number of bits) from processor to processor. A word indicates how many contiguous bits a processor can interpret and act on at any one point in time. When a processor fetches the next instruction or item of data from main memory, the word length indicates the maximum number of bits that it might transfer.

The word length of a processor confers two important characteristics on its host system. First, any instruction (and, in some cases, the address of the memory locations whose contents are to be operated on) to be executed by the processor is linked to the word length. Thus in general, a longer word length indicates a more powerful processor. The instructions that the processor is able to obey directly are known as its *instruction set*. In many ways it is the fundamental parameter of a processor. A larger number of bits available for coding the instructions in general implies a larger, richer, more powerful instruction set, and thus a more powerful processor.

Second, it is important to realise that our discussion of storage and processing above only involved information represented as characters. Obviously, in data processing we are primarily concerned with storage and processing numbers as numbers, not as a group of characters. It is crucial to appreciate that the number 67.4 is fundamentally different from the individual characters '6', '7', '.' and '4', which happen to be stored in that sequence. It is feasible to store and process numbers represented internally in the machine as strings of characters such as this, but it is extremely wasteful of memory capacity. It also makes for a very, very slow computer, since each character is the subject of an individual fetch from memory; and makes arithmetic difficult. Thus, within most computers there are in fact several different storage formats, one of which is, of course, character form. The other forms are largely concerned with storing various numeric quantities — either *integer* (i.e. whole numbers) or *floating point* (i.e. with a decimal point) — to varying degrees of accuracy.

To store a floating point number takes at least 32 bits; thus personal computers (typically 8-bit) and minicomputers (typically 16-bit) take more than one instruction to manipulate these numbers. Most large business machines have 32-bit processors, which speeds up the processing of decimal arithmetic; but for really complex and accurate scientific calculation, 64-bit processors have been developed.

4

Software – Instructing the Computer

In the previous chapter we described the major pieces of equipment that go to make up a computer system, and illustrated the technical basis on which they operate. In this chapter we describe the various software components and illustrate their function. The first point to bear in mind is that software, as its name suggests, is somewhat less tangible than the pieces of equipment that we have just described. Software comprises the instructions to the equipment which enable it to function usefully from your point of view — they are only tangible when printed or displayed in some human-readable form. As we shall see, a set of instructions may exist in one of several forms within the machine, each of which requires different skills for their understanding. For the purposes of describing its function, software is usually divided into two distinctive categories: *operating system software*, which includes the instructions necessary to enable our equipment actually to function as a computer system; and *application software* or programs, which instruct the system on how to perform a particular user-oriented task.

In the first part of this chapter we will concentrate on what is known as application software, or programs — instructions specifying the necessary steps to perform a user-oriented task such as preparing a payroll, updating a file of stock records, preparing an invoice, and so on. A complete set of instructions to perform some particular task is known as a *program.*

The Nature of Software

All programs have three basic components namely, *input*, *processing* and *output*. First, nearly all programs need some input data on which to work; for example, this might be the weekly time sheets, a batch of invoices, a series of cash flows, a discount rate etc. Accurate data input is a vital part of data processing and we discuss it more fully in Chapter 9. Second, the input data is processed — multiplying hours worked by rate per hour, cash flows by discount rate etc. Processing may also involve the retrieval of previously stored information; for example, to get pay to date, tax to date, etc. And, increasingly, it may also involve communication with another computer or remote terminal. Finally, after the relevant processing has been done, then the results must be displayed or reported in some form, such as payslip, the sales ledger or simply a net present value. Thus a simple model of any program could be shown as in Figure 4.1.

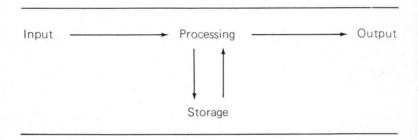

Figure 4.1 *A Simple Model of Data Processing*

To understand a little more about the nature of software, it may be useful to consider the analogy of making a cake. Consider the instructions in Figure 4.2. The inputs would be butter, eggs, flour and sugar, the processing instructions might be: 'Mix ingredients together, put in greased tin, put tin in the oven (350°F), remove tin from oven after 20 minutes'; and the output would hopefully be a sponge cake. In this example a number of separate inputs have been combined and manipulated to produce the desired result. The processing instructions contained a number of key words that were meaningful to the

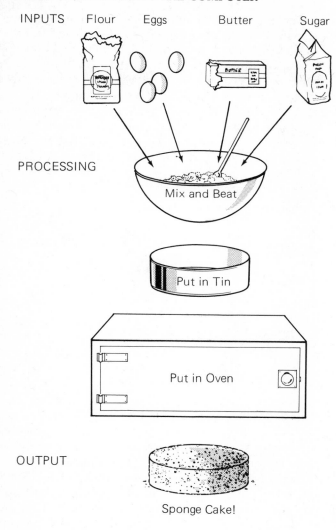

Figure 4.2 *A Cooking Analogy*

'cooking processor'. However the analogy with a computer program breaks down, because with the computer there is no fundamental difference between the data (input and outputs) and the processing instructions themselves.

Consider next the simple data processing example shown in Figure 4.3, where the inputs are a series of names; Jones, Brown, Smith and Williams. The processing instructions are to sort these names into ascending alphabetical order. Note that in this context 'sort' is a keyword which has a special meaning to the computer. The output is then a list of names sorted in alphabetical sequence: Brown, Jones, Smith, Williams. This example illustrates the point that, in a data processing environment, the 'data' to be processed need not be numeric information; in this example it was characters (often referred to as *text*).

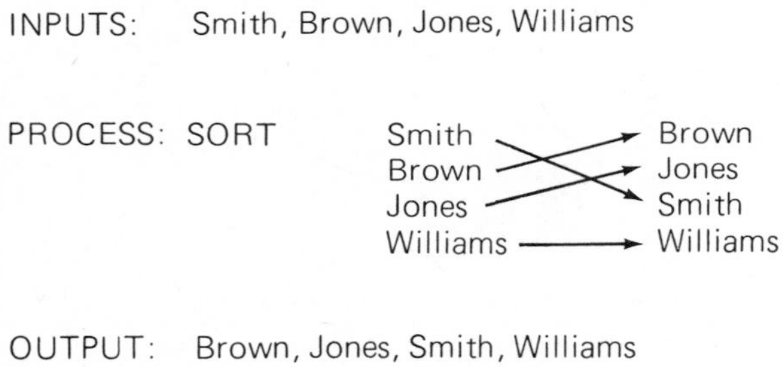

INPUTS: Smith, Brown, Jones, Williams

PROCESS: SORT   Smith ⟶ Brown
                   Brown ⟶ Jones
                   Jones ⟶ Smith
                   Williams ⟶ Williams

OUTPUT: Brown, Jones, Smith, Williams

Figure 4.3 *Sort Example*

The input data will be represented as a series of bits (binary data), as will the processing instructions. Consequently, it is not uncommon for the data being processed by one program to be the instructions for another program. Later in this chapter we will describe the role of an operating system, which is the major component of system software. Whereas application software is specific to one particular user task (such as payroll, or stock recording), system software is generic to all application programs and is shared by all users.

All software specifies the logic — the set of rules — by which the 'data' are to be handled. Note that, in the context of this book, we are concerned with the processing of all information — be it in data, text, voice, video or facsimile form — and that the operations specified in this chapter, though nominally written as 'data processing', could be applied to all forms of information

processing. We can summarise the basic building blocks of any
system as instructions to the hardware to:

   accept   *store*   *retrieve*   *process*   *transmit*   *display*

data (or information). All of the above steps will be present to a
greater or lesser degree in any application program. These steps
specify what the application program actually causes the
computer to do with the data — it will accept, store, retrieve,
process, transmit and display the data on receipt of the appropriate instructions. But what form do these instructions take?
Why can't we simply instruct the computer in straightforward
English?

Machine Code Programming

All instructions and data internal to the computer must be coded
in some agreed fashion, using the binary number system.
Furthermore, the processor at the heart of the computer system
can function only on receipt of instructions drawn from those
available in its own repertoire, known as its *instruction set*. Thus
despite the way in which we as humans may arrive at the
function and form of the instructions that we wish the computer
to obey, the processor dictates a fundamental requirement — the
instructions that it obeys must be presented as binary codes
according to the specification of its instruction set.

The programmers of the first computers back in the 1950s had
no choice — they had to program directly in binary using the
processor's instruction set — in what today is usually termed
machine code. This procedure had several major disadvantages,
such as:

> it was relatively laborious, since machine code instructions
> are very simple;
>
> it was very complex, as the programmer had to have a good
> knowledge of the workings of the computer;
>
> as everything was in binary, it was easy to make mistakes,
> but difficult to find them (and almost impossible to find
> them in other people's programs).

Thus there began a process to try to reduce the impact of these
deficiencies, and ultimately to increase the productivity of the

programmer. The first stage was to replace the binary codes of the instruction set with an exactly comparable set of mnemonics, such that there was a unique one-to-one correspondence between each mnemonic and each instruction in the processor's set. These mnemonics were thus still at the detailed level of the processor's instruction set — for example, ADD to add together two numbers, MOVE to move a number from primary memory to a specified i/o device, and so on. The advantages from a programmer's point of view were that they were less likely to make mistakes with the mnemonics and the resulting program was easier to follow and comprehend.

Even so, programming in these mnemonics — usually known as *Assembler* — was still very tedious and prone to error. For example, an Assembler program to add three numbers together might be written as follows:

memory location (address) :	19	20	21	22
contents (data) :	125.2	73.1	527.8	

i.e. the three memory locations (pigeon holes) numbered 19 to 21 inclusive contain the numbers 125.2, 73.1 and 527.8 respectively.

We will assume that our processor in this example has two reserved memory locations (called *registers*) that its arithmetic unit accesses when adding; these will be referenced as registers A and B. Our program might then be:

1	ZERO,B	(set contents of register B to zero)
3	MOVE,A,19	(move the contents of memory location 19 to register A)
5	ADD,B,A,	(add the contents of register A to those of register B; result is left in B)
7	MOVE,A,20	(move contents of location 20 to A)
9	ADD,B,A	(add contents of A to those in B)
11	MOVE,A,21	(move contents of location 21 to A)
13	ADD,B,A	(add contents of A to those in B)
15	MOVE,22,B	(move contents of register B — the answer — to memory location 22)
17	END	(no more instructions)

These instructions are executed in sequence until the 'END' statement. (There could have been an instruction to transfer control to a point in the sequence other than the next instruction.) The numbers to the left of the instructions represent the memory locations where they are stored (two locations for each instruction). So that in primary memory, locations 1–18 represent the instructions and locations 19–22 represent the data. If the 'END' instruction was missing, then the computer would have attempted to interpret the data in location 19 as an instruction, which would probably have caused it serious indigestion.

Apart from being laborious to develop, machine code programs have one other great limitation. They can be used only on the specific make of computer for which they were developed, thus 'locking-in' the user to a particular manufacturer. The advantage of machine code is in its efficient execution — this, however, is becoming less critical with the increasing power of the hardware.

High-Level Programming Languages

To overcome some of these problems high-level languages were developed. These went some way towards improving the productivity of programmers, and enabled the programs to be moved between different manufacturers' computers (albeit, with difficulty). We still haven't answered the question — why can't we instruct the computer in plain English? The problem is that in general usage English is ambiguous — it is often the case that a long and complex statement (such as one that might be necessary to instruct a computer on how to perform some realistic application) is capable of being interpreted in more than one way. It may be clear to a human reader of these instructions how to interpret them in the light of the specific context in which they are set. But a computer is in general incapable of such sophistication, and is unaware of the specific context of a given set of instructions. The instructions presented to the computer must be unambiguous in their interpretation. A high-level programming language is a very restricted subset of the English language, in which each word has a unique and precisely defined meaning.

The first high-level languages were developed in the second half of the 1950s. The first in general use was *Fortran* (which stands for FORmula TRANslator) in 1957, for which the first international standard was developed in 1959. As its name suggests, Fortran is a language designed with scientific applications in mind. Thus Fortran can easily handle complex mathematical formulae, but its data processing capabilities are relatively weak (since scientific applications traditionally did not involve substantial volumes of i/o).

However, typically commercial data processing applications do involve considerable amounts of i/o, whilst the computations to be performed are often trivial. Thus *Cobol* (Common Business Orientated Language) was developed by 1961, though it did not receive significant widespread support from the computer manufacturers until 1966, when the US Department of Defense specified the Cobol language as a requirement in a tender for 100 computer systems. From then on Cobol achieved widespread acceptance, and today it is by far the most commonly used language (estimates suggest that roughly 70% of all application programs are written in Cobol). Together with RPG, PL1 and Algol, Fortran and Cobol are the languages usually to be found in large commercial data processing centres.

The development of interactive, time-sharing computer systems in the 1960s, and the subsequent development of cheap personal computers in the 1970s has led to the popularity of the *Basic* language (Beginners All-purpose Symbolic Instruction Code). Basic was specifically developed for use from terminals, having a minimum of formal i/o conventions, and was intended for use by non-computer personnel. This latter point has meant that it is by far the easiest computer programming language to learn, and is typically the one taught to school children and many students. Unfortunately, its popularity has led to a proliferation of alternative versions, so that today Basic is one of the few languages without an internationally-defined standard.

Characteristics of the Major High-Level Languages

In a high-level language each program statement (i.e. each instruction in the high-level language) is equivalent to many machine code instructions (hence the 'high-level' designation). The difference in number of statements between a high-level language and machine code could easily be 10-fold (i.e. there would be 10 times fewer statements in the high-level language version). In addition, the memory locations to be addressed by these statements could now be designated by user selected names, rather than by their absolute binary addresses as required when programming in Assembler. From the programmer's point of view the advantages were considerable — far fewer instructions to write (and 'debug'), and much more easily readable and maintainable code.

We will briefly review the characteristics of the three most commonly found high-level programming languages, namely, Fortran, Cobol and Basic.

Fortran

Fortran was the first of the high-level languages to be developed and, as its name suggests, it was primarily intended for use in applications involving mathematical procedures. Thus it continues to find its major use in scientific and engineering applications. However, because such applications characteristically involve complex computations, but relatively small volumes of input/output, Fortran's ability to represent complex mathematics or logic is high, whilst its input/output capabilities are limited and rather arbitrary. Various international standards for the language have been agreed upon. The most recent is Fortran 77, though its predecessor Fortran IV is more widely used.

As an example of a Fortran statement, consider the following:

$$NETPAY = (HOURS*RATE) - TAX$$

whose meaning is probably fairly clear to you. Two *variables*, HOURS and RATE are multiplied together (when instructing a computer the multiplication symbol is '*'), and then the variable TAX is subtracted from this product to give the value of the variable NETPAY. All high-level languages use *symbolic variable names* (e.g. HOURS, RATE, etc), chosen by the programmer, to reference memory locations, and to express the operations required in their contents. That is, the computer will assign a symbolic name used by the programmer to a unique actual memory location which it will automatically access every time that particular name is used. Thus the programmer can program using symbolic names that have meaning in the context of a particular program. Consider a second Fortran example, this time analogous to our previous machine code example. Suppose that as before we wish to add three numbers together, then a Fortran program might be:

$$TOTAL = 0$$
$$DO\ 100\ J = 1,3$$
$$100\ \ TOTAL = TOTAL + RENT(J)$$

The 'DO' statement tells the processor to execute the statements up to and including that labelled '100' for each value of the index variable J as it increments from its initial value (in this case 1), to its final value (in this case 3), in steps of 1 (which is implied in the form of the statement). Thus, in the example, the statement labelled 100 is executed three times, with J taking on the values 1, 2 and 3 successively, and thus accessing the three memory locations occupied by the variable $RENT$. This illustrates a further point of advantage when programming in a high-level language: namely, that a symbolic name can be given not only to a single memory location, but also to a consecutive block of such locations, considerably aiding the processing of large tables of data. In this example the variable $RENT$ refers to such a block of memory — the specific individual location to be accessed within the block is indicated by the subscript J. This example is a very small illustration of the greatly increased power available to the programmer when using a high-level language. It is not that it enables him or her to do anything over and above what can be done using Assembler, the point is that it can be accomplished so much more speedily and accurately. High-level languages primarily improve the productivity of the programmer.

Cobol

In contrast to Fortran, Cobol was explicitly designed to be a language whose syntax would be especially suited to applications in commercial data processing. Relative to Fortran, statements in Cobol are in one sense at a higher level in that they are closer to pure English. Indeed, it was one of the design objectives of Cobol that the basic purpose of a program should be understandable from reading its statements. Because Cobol is closer to English, the process of translation into machine code is much greater relative to that of Fortran. As in the case of Fortran, well-established international standards for the implementation of Cobol have been established, the current one, and the one most generally implemented by computer manufacturers, being Cobol 74.

The capabilities of Cobol are much more oriented towards the handling of large volumes of input/output than are those of Fortran, especially in the handling of files of data and in the formatting and layout of output, but are much less oriented towards handling complex mathematical procedures. For example, the first of our simple Fortran examples above would be written in Cobol as follows:

MULTIPLY HOURS BY RATE GIVING GROSS PAY

SUBTRACT TAX FROM GROSS PAY GIVING NET PAY

As can readily be seen, this is very close to English and its function can be easily understood, even if a little verbose! Another feature of Cobol is that the program is divided into four different sections, or divisions. Each division has a different function — for example, to describe the operating environment of the particular computer. The specification of the file and data layouts is separated from the processing instructions; so that, for example, the less skilled programmers could simply code up the processing instructions and not have any control over the file and data specifications.

Basic

Basic is a more recent development in programming languages, but is not significantly 'higher' than Fortran or Cobol. Rather it simply had a different objective in its design. Basic was originally designed as an easy-to-learn and use language, particularly oriented towards problem-solving programs developed on interactive, time-sharing computer systems. Consequently, the specification of the language was kept as simple as possible, and in particular the input/output conventions were kept to a minimum. For these reasons, Basic has proved to be a very popular language in which to get started on programming, as it enables relatively inexperienced programmers to produce programs handling complex problems in a fairly short period of time. However, in terms of encouraging good programming practice, Basic leaves a lot to be desired (largely because it sacrifices formal structure in the interests of ease of use), and is generally considered to be unsuitable for use in large-scale data processing applications. Also, versions of Basic have proliferated over the years so that now there is little common standard amongst the various implementations of the language.

As an example of the use of Basic, the example presented previously when discussing Fortran could be written in Basic as follows:

$$P = (H^*R) - T$$

The similarity of the above with the Fortran equivalent can be seen; in general Basic only allows single character (or character plus a number) as variable names, making large Basic programs difficult to read and understand. However it must be added that Basic retains great popularity amongst the amateur programmers and users. It has become the *lingua franca* of the microcomputer and there are probably more people in the world who can program in Basic than in any other language. It is rather scorned by the computer professionals and this may be in part because it threatens their existence by enabling the end-user to instruct the computer.

Structured Programming

Programming a computer, even in high-level languages, has always been an art. It has a strong creative element and ten programmers given a problem will come up with ten different (often radically different) programs. They will all achieve the same output for a given input, but they will probably behave differently when operated outside their original design specification and be more or less difficult to modify. The difference in performance amongst the programmers may be astonishing. Tests (at Boeing) on a large sample of programmers given the same problem have shown productivity ratios of 5 to 1 between the best and worst programmers. It is not at all obvious who are the best and worst, due to the complex interactions that go on in developing big systems.

Structured programming is an attempt to turn software production into a science. Hopefully, using this technique there will be only one way to write a given program which will allow for good quality control and better maintenance. However, before this is possible, some extensions have to be made to the high-level languages; thus, we now have Structured Cobol, Fortran, Basic, etc. Alternatively a new language can be developed for this specific purpose. *PASCAL* is such a language which is gaining great popularity amongst computer professionals in America, particularly on college campuses. A new language called *ADA* is being designed to supersede even this but it is possible that procedural languages may be eclipsed altogether as we discuss in the next section.

Non-Procedural Software

All the software we have discussed up to now has been procedural, in other words it is concerned with *how* to produce what is required. All of the various high-level languages already discussed are procedural in nature — they provide detailed instructions on how to perform a specific task. A more desirable approach is to specify *what* is required and let the computer figure out how. This is a lot more difficult than it sounds. For some time it has been possible to do this within the context of a particular application but new tools are becoming available which allow standard dp development to be done without writing a single line of procedural code.

A good example of this kind is the RAPID/3000 software available for the Hewlett Packard 3000 minicomputer, as illustrated in Figure 4.4. All inputs of data can be specified through a system called VPLUS/3000. Using this system, the screen layouts, all validation rules and data items can be specified interactively. Secondly, the data to be stored can be referenced and specified in a data dictionary, and the relationships between items can be defined using the IMAGE database management system. Thirdly, by employing a transaction processing system, TRANSACT/3000, the transaction processing rules can be specified in a task-related way that interfaces with the IMAGE database through the data dictionary and with the user via VPLUS. And finally, *ad hoc* enquiries and regular reports can be

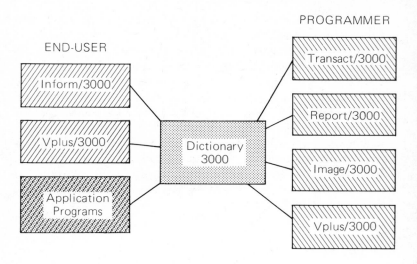

Figure 4.4 *Hewlett Packard RAPID/3000 System*

generated by systems such as QUERY, REPORT and INFORM.

Thus, using such software tools, a large dp system could be developed which does not include a single line of traditional procedural code. It is claimed that these types of tools can give productivity gains of 10 to 1 compared with the typical gains from structured programs of 1.5/2 to 1. The demise of the Cobol programmer has long been predicted, but we can now see some of the practical tools which will hasten this process.

Programmer versus User Productivity

One of the great motivations behind the development of high-level languages was to increase the productivity of application programmers. Initially this was successful, but in recent years the emphasis has moved towards increasing the productivity of the end-user, especially with the development of on-line inter-active systems tending to remove the buffers between the users and their applications. Also the continuing reductions in the cost of computer resources mean that application designers can be less concerned with the efficiency of their systems from a computer point of view, and more concerned with their effectiveness from a user point of view.

One of the factors stimulating such trends is the growing availability of application software systems, providing a general capability to a (professional) user in a particular area, rather than a specific application program, solving a particular problem. For example, application software systems are being developed for engineering designers in specific areas (e.g. cars, bridges and other structures, and electronic circuits), in which the engineer can directly, usefully and meaningfully communicate with the computer system, using his own terms and concepts. These are commonly known as *application languages*. Other examples include languages for project planning, process control engineering and financial modelling.

System Software

It is crucial to appreciate that in the absence of any instructions, our computer equipment would not function at all. This is because what we are primarily describing in this book is, strictly speaking, termed a general-purpose, stored-program, digital electronic computer system. It is a general-purpose machine — it can tackle any problems for which we can provide an appropriate set of programs and data. This can be contrasted with other computers which are dedicated to a single specific task — perhaps the best example is seen in microprocessors devoted to the control of some particular item of equipment, such as a washing machine, digital watch, electronic game, and so on. These microprocessors are computers and they are stored-program; but the program is included at the time of manufacture and subsequently cannot be modified or changed.

However, the general-purpose stored-program machine with which we are concerned requires instructions before it can function. If we switched on the power to our items of computer equipment, but did not provide any instructions on how to operate them, nothing would happen, except that some electrical energy would be consumed and some heat given off. But you may be saying, 'I've seen computer systems which do become active in a user sense as soon as the power is switched on to them'. True, but what is happening in such situations is that the processor automatically begins to execute instructions located in a specific area of the computer's primary memory (in a ROM) which itself loads and executes other programs, which in turn bring up a running system for the user. A procedure such as this is known as *bootstrapping*.

Operating System

Until the mid-1960s computers came with a series of specific applications and utility programs. Each new program had to concern itself with the effect of interaction with other programs and how to store data on the filing system.

When IBM introduced their 360 range it introduced a new concept in organising the computer. The IBM 360 came with an operating system (DOS/OS 360), which was a program that managed the operating environment, thus taking that requirement away from each application program. Quite rapidly all the other computer manufacturers adopted the same approach. *Operating systems, executives, supervisors,* and *monitors*, as they are variously called, became an essential element of any computer system. Even today's low cost personal computers have them, for example, Apple DOS (Disk Operating System).

There are at least three major components of any operating system.

Controllers

First, there is the basic set of instructions for each individual piece of equipment, instructing it on how to function as a useful component in the overall system; for example, instructions on how to take a block of binary data from the processor and to generate an appropriate pattern of bits on a magnetic tape to represent that data; similarly for a magnetic disk, but in this context movement of the read/write head relative to a particular track, the sensing of the beginning of that track, and initiation of the magnetic writing process have also to be accomplished. The software and hardware to accomplish tasks such as these are commonly referred to as *drivers* or *controllers* — they drive or control the appropriate piece of equipment.

Compilers

The second major component of an operating system is the software necessary to interpret the application program. As we have previously illustrated, ultimately the instructions that the processor is asked to obey must be drawn from its instruction set. But its instruction set only exists as a set of binary codes, and whilst it is feasible to program in machine code it is not usually desirable.

In order to be executed by the processor, the high-level language statements have to be translated into machine code. This is done by processing the original language statements (known as the *source code*) through a *translator* program, known as a compiler, to produce as output a machine code version of the original source code program (known as *object code*). This process is shown in Figure 4.5. It is important to realise that a compiler is a program just like any other — in this case, its input data is the source code statements in a high-level language, and its output is the corresponding object code. This translation process, or compilation is undertaken only once (assuming that the source code is correct). The program is run by having the processor execute the object code as and when necessary.

However, a compiler is a general-purpose program, designed to translate any valid source code in the relevant language into object code. For any non-trivial, practical piece of source code there are usually several alternative equivalent sets of object code possible. That is, there is no unique translation from source code to object code. For this reason, the object code resulting from a compilation may be relatively inefficient, as compared to what would be possible if the problem had been coded directly in machine code. (Inefficiency is usually measured in terms of cpu time taken to execute the particular piece of code.)

In some applications, where fast execution is paramount (typically on-line, real-time systems, such as airline reservation systems), coding will be done directly in Assembler to ensure as efficient a code as possible. A different compiler is needed for each high-level language to be implemented on a particular machine — a compiler is tied to one particular machine (or more specifically, to one particular instruction set). However, a high-level language is potentially more general, since a program

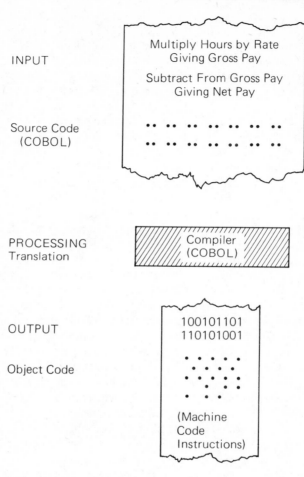

INPUT

Source Code
(COBOL)

PROCESSING
Translation

OUTPUT

Object Code

Figure 4.5 *The Compilation Process*

written in such a language can theoretically be run on any machine on which a compiler is available for that language.

However, unfortunately theory isn't quite borne out in practice. International standards have been set down for almost all of the major high-level languages, usually by a specific committee typically with computer manufacturers comprising a majority of the members. But in this situation, agreement is a long-drawn-out process, and even when it is achieved, the specification of the standard tends to the lowest common denominator of the implementations of the language in question by the various computer manufacturers (since no manufacturer wants to agree to a language specification that it can't meet itself). However, all computer manufacturers have some specific design characteristics of their computer systems on which they sell their systems, and which they want their users to be able to exploit. Generally the way to do this is to provide some specific extensions (specific that is to that manufacturer) to the current language standards. The users (systems analysts and designers) naturally wish to exploit the special features of the computer system (probably one of the major reasons for the choice of this system), and therefore will make use of the language extensions. Thus the application programs will be written in a non-standard version of the language. Even without this situation the specifications of the language standards are themselves on occasion not unambiguous, so that implementation by different manufacturers often differs.

Operational Control

The final component of the operating system is a set of programs designed to allow the system to operate itself in as automatic a way as possible. Given that our machine is highly likely to be capable of processing hundreds of thousands, if not millions of instructions per second, it is clearly not a particularly effective use of this resource if, at every point that a decision is required in scheduling its resources, the machine has to wait for an operator response. Thus in present day computer systems the role of operator is largely confined to a few purely mechanical tasks which are difficult to automate, such as mounting a magnetic tape, or taking paper off the lineprinter. Otherwise the system itself manages its own operation, scheduling its resources to meet the tasks imposed on it, with the objective of maximising the utilisation of the cpu on application programs.

Modes of Computer Operation

Batch Processing

One of the great problems in achieving this objective is that the various hardware components of a computer system function at very different speeds, in terms of the rate at which they can process, locate and transfer instructions and data. The cpu itself operates by far the fastest, and the input/output peripherals the slowest — especially keyboards with human operators! In the first operational computer systems in the 1950s, the system worked on one job at a time, determined by the program being run. All the transactions pertaining to a particular processing task would be batched together over a period of days, then processed by the program as a single batch. Once one program had finished processing, the next would be loaded into memory and executed with its batch of transactions (orders, invoices, time sheets, etc). This form of organisation of the operations of a computer system is known as *batch processing*, and in some form or other is still the dominant way of organising processing in computer systems today.

On these early systems the initiation of each batch run was done manually by the operator, and all peripheral devices (card-readers, printers, etc.) were directly connected to the cpu. Needless to say overall utilisation of the cpu was extremely low! It spent most of its time waiting, either for input or output to a slower device to be completed, or for the operator to initiate the next job. These problems were alleviated by breaking the direct connection between the cpu and input/output peripherals. All input would be placed on a (tape) file, as would all output; the cpu can transfer data to and from tape at a much higher rate than from a card-reader or to a printer. Thus, a whole stream of jobs would be built up on an input tape from several input devices, and similarly the output from these jobs would be placed on an output tape to be printed when appropriate. The actual reading of cards or printing would go on independent of the cpu. At the same time, initiation of the next job would be automatic as the next job on the input file. This process is known as *spooling*. It significantly improved the throughput of work and increased the utilisation of the cpu.

Multiprogramming

Although spooling enabled the speeds of the cpu and input/output operations to be more closely matched, the cpu was still only working on one job at a time and could still be idle for relatively very long periods of time. This is because most computer processing, especially that of a commercial nature, involves large data files stored in secondary memory. A fundamental part of the processing of such data is the necessity to transfer blocks of it to and from primary memory for processing. In the majority of commercial applications the time spent accomplishing these data transfers far exceeds the time actually spent in processing the data. Thus the next stage in the development of operating systems was to enable the processor to switch to a second or subsequent program as soon as it encountered a wait condition on the program which it was currently executing. Typically, a 'wait' condition would occur when the program required the transfer of another block of data to or from memory. This process is illustrated in Figure 4.6.

Thus the processor would commence executing program 1, the first in the input queue; if, as is likely, this program requires a file access, its execution is suspended whilst the access is executed by a separate controller, and the cpu transfers to program 2 and begins executing this program; if program 2 now requests a file access it too is suspended, the operating system checks if the file access for program 1 is complete, and if so resumes its execution; if not, it commences to execute program 3; and so on. This process is known as *multiprogramming*. Note that the operating system has to have the capability of suspending execution of a program by the cpu and being able to return and resume execution after executing some other job.

In this form of multiprogramming, execution can be transferred between programs only at a 'natural break' in the execution of a program. However, it might be the case that in the above example, program 1 had a very high priority such that, once the requested file access was complete, we would want the operating system to halt execution of program 2 and resume that of program 1 immediately, rather than waiting for a 'natural break' in program 2 as at present. This requires that the operating system have the capability to *interrupt* an executing program and transfer control to another program. Such a capability is

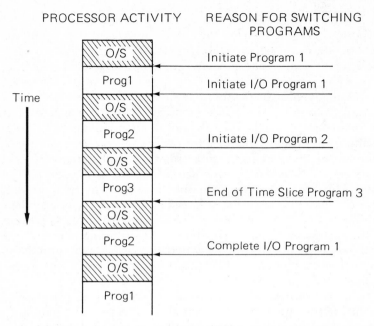

(a) Processor activity over time (O/S indicates operating
    system overhead switching between programs)

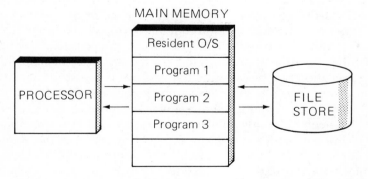

(b) Allocation of main memory in multiprogramming
    environment

Figure 4.6 *Operation of a Multiprogramming System*

necessary if a priority structure is to be followed in processing jobs in a multiprogramming environment. However, an interrupt capability is crucial to the development of on-line, interactive systems.

On-line, Interactive, Real-Time Systems

In the above methods of operating system organisation, the input/output peripherals are (deliberately) not directly connected to the cpu — they are spooled, off-line devices. One of the consequences is that no possibility exists for direct communication between the user (or programmer) and the executing program whilst it is actually being executed. It is increasingly the case that the design of more flexible, and user-oriented systems requires that they be designed with on-line, interactive implementation in mind. For example, on-line data entry by the originators (or 'owners') of the data can greatly contribute to overall data accuracy (and identifying and correcting data inaccuracies once some processing has been undertaken is often a difficult, time-consuming and expensive operation); most forms of computer-based modelling are significantly more effective if the modeller can interact directly with the system; systems which are designed to communicate directly with the end-user, such as automatic bank-teller machines, obviously require such a capability.

The term *on-line* indicates that the user can directly submit work to the computer and receive the results without any human intermediary. If the user has no opportunity to interact with the program, then this mode of operation is known as *remote job entry*. However, if the user can influence the program once it starts to execute by entering data, selecting options, specifying the results required, etc, then that is called *interactive computing*. Obviously interactive systems must be on-line. The difference between this approach and that of batch processing is illustrated in Figure 4.7.

The term *real-time* indicates that the system will respond to some transaction within a given (short) time, and that this response time is a critical performance parameter of the system. Typically these systems will only be running a single application and be optimised to support a large number of terminals. This term is also used in process control systems to indicate that the

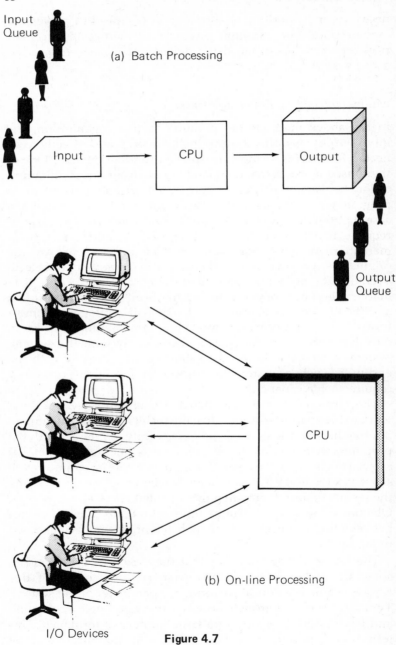

Input Queue

(a) Batch Processing

Input

CPU

Output

Output Queue

(b) On-line Processing

I/O Devices

Figure 4.7

system must respond to a given signal within a critical time, e.g. shutting a valve.

Time-Sharing Systems

Finally, there are two further approaches to the design of operating systems which deserve a mention before concluding this section. Firstly the development of generalised time-sharing systems at the end of the 1960s in response to a requirement for a system that would support a number of users running a different interactive application program simultaneously. These systems grew up around new designs in hardware, particularly mini-computers.

In a *time-sharing system* all users are connected on-line, typically via relatively low-speed input/output devices, to a cpu which shares its time between them in a 'round-robin' fashion. Thus user 1 gets, say, 20 milliseconds of cpu time, followed by the same for user 2, user 3, and so on in turn back to user 1, who, hopefully, hasn't noticed the interruption! Each user apparently has personal use of the whole computer (it soon becomes evident that this is not the case when the number of users causes the response time to deteriorate). In essence it's another approach to matching a high-speed resource with many slower-speed devices. Time-sharing systems grew in popularity and availability throughout the 1970s.

'Virtual' Systems

The second, perhaps much more fundamental development in operating systems is the concept of 'virtual memory', and indeed of a 'virtual machine'. One of the problems encountered in developing the multiprogramming operating systems described above was that they required that all of the programs (and their associated data) be in primary memory. In practice, primary memory in such systems was divided into a small number of partitions (initially of fixed size, but in later versions variable), with different priorities on each partition, and in effect separate job queues for each partition. One of the problems that arises from this approach is that it demands ever increasing amounts of (relatively expensive) primary memory. The virtual memory concept involves breaking both the applications program and its data into small *segments* or *pages*, and keeping in primary memory only those pages currently being used (code segments can be shared between several users). All other pages are kept in high-speed secondary memory (originally a drum, but now fast disk). Another component of the operating system, the memory manager, has to decide which pages to keep in main memory, on the basis of frequency of use, next in sequence, priority structures, for example. In some systems, a large part of the operating system is also paged in this way, with only the core being permanently resident in primary memory.

At some point the operating system will discover that a page it needs in order to continue executing a program is not in primary memory and will have to be copied in from secondary memory, and similarly for data. If there is space available in primary memory, then there is clearly no problem. However, the general situation is that something will have to be moved to make room (in practice, code segments can be over-written, in that they are not changed by processing, whereas data must be copied out first). This process is known as *swapping* (or *paging*). It was made available on an operational system in the early 1960s (on the Ferranti Atlas), but has become widespread practice only in recent years. An example of this process is show in Figure 4.8.

In effect, a multiprogramming, on-line, interactive, virtual memory operating system is the culmination of all the developments in operating system design described above. In general, it is much better at allocating primary memory than a partitioned multiprogramming system, and can offer a much more flexible system to the user than a general time-sharing system. To become effective in commercial systems, it needed the availability of large primary memories and fast, reliable disk drives. Most new operating systems are based on some form of implementation of this concept.

System Performance

It should be borne in mind that all of these improvements in operating system effectiveness and efficiency are not without their cost. An operating system is a set of instructions, a program just like any other, albeit usually somewhat larger and more complex. But it too requires system resources, processor time

MAIN MEMORY   SECONDARY MEMORY

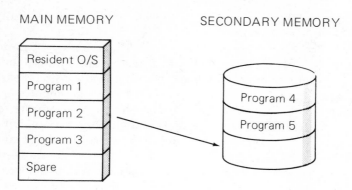

Operating systems require Program 4 for execution which is currently not resident in main memory. There is insufficient spare space to load Program 4, therefore some resident program(s) must be swapped onto secondary memory.

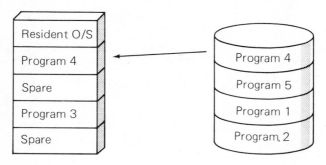

Figure 4.8 *Illustration of Swapping*

and memory, for its execution — and it is running all of the time. Its effectiveness and efficiency will have a direct and unavoidable impact on all user applications. As operating systems have become more and more sophisticated and complex, so their demands on system resources have become greater and greater. It is not uncommon in large multiprogramming systems for the system software physically to take up somewhere between a third and a half of primary memory, and to consume as much as half of total cpu time. Thus the throughput of a system, as measured in terms of productive user work, may well be much less than might have been thought from the hardware specifications of the system.

In this context, there have been major differences in operating system efficiency between a typical mainframe system and a minicomputer system. A typical mainframe operating system has gone through most, if not all, of the development stages mentioned above, with the result that often it now appears as a large, rather unwieldy beast, valiantly attempting to be all things to all users. In contrast, minicomputers developed out of small real-time control systems used for process-control, instrumentation and weapons-control and were designed from the outset as on-line, interactive systems which to a significant extent could be tailored to the individual characteristics of each user's environment.

5

File Organisation and Processing

In this chapter we will examine how the processing of transactions may be organised, and how the files of data associated with this processing are structured and managed.

File Organisation

An electronic file in computer technology is exactly analogous to a paper-file file in an office context. It is a collection of logically-related material grouped under a single heading — for example, a personnel file, a customer file, a stock file, and so on. A file consists of a number of records or entries, each of which contains all of the data for each entity in the file. Thus, in the personnel file, there would be a record for each employee; in the customer file, one for each customer; in the stock file, one for each product; and so on. A record contains the data for that entity as a set of fields or data-items. Thus, each personnel record might have as data items works number, employee name, address, hourly wage rate, tax code, deductions code, bank sort code, bank account number, and so on. Similarly, each record in the customer file may contain a customer code, the customer name, invoice address, delivery address, credit rating, contact name, and so on.

The hierarchy of data storage is thus as follows:

database is a collection of related files,

files are collections of similar records, organised in some sequence,

records are collections of logically related data-items, to be treated as a single unit,

data-items are the individually respresentative items of data,

bits/bytes respresent the characters and numbers in the data-items.

An example of the data hierarchy for a sales order system is shown in Figure 5.1.

Such files may of course be very large — the personnel file for a large company may have several tens of thousands of records, whilst the stock file may have hundreds of thousands of records, most of which are active each day. As we have seen earlier, such files have to be stored in secondary memory, and records (either in blocks or individually) retrieved and copied to main memory for processing. For large files considerable care and effort has to

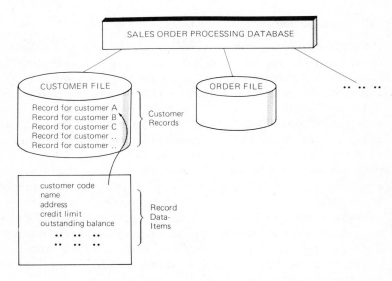

Figure 5.1 *Example of Data Hierarchy*

be given to the logical organisation of such files, the way in which they are to be stored in secondary memory, and the way in which transactions pertaining to the contents of the file are to be processed. Although we will be discussing these points as separate items, it is important to recognise that in practice they are highly interrelated and the decisions about file organisation, storage and processing are joint ones.

In practice there are two main alternative ways of organising and processing data files — either sequentially, or using some form of direct access. Each method has relative economic and technical advantages and disadvantages, depending upon the needs and characteristics of the specific application. For example, characteristics having a significant impact on the choice of file organisation and processing method are:

> *response time*, defined as the time which elapses between the request for processing and its completion; in some applications, response time is a critical parameter.

> *activity rate* on the file; that is, within a given time, what proportion of total records in a file are accessed; the economics of the processing methods are significantly affected by this rate.

> *volatility* measures the rate at which records are added to or deleted from the file; again, this can have a significant impact on the economics of alternative processing methods.

> the *size* of the file, measured either by the number of records or the total number of characters in the file, can have a major impact on the type and capacity of the physical media used to store the file.

> and finally, the degree of *interdependence* and *sharing* of a file can have a major impact on both organisation and processing; files used by several different applications have to be organised and managed in some commonly agreed manner, and may well lead to their implementation using some *database management system (DBMS)*.

It should be recognised that the basic record management operations on a file are common to all files, irrespective of their organisation. For example, in general all files will require:

procedures to *create* a new record;

procedures to *delete* an existing record;

procedures to *retrieve* the contents of a specific record;

procedures to *update* the contents of a specific record;

procedures for *backing-up or archiving* some or all records, and corresponding procedures to *restore* records from an archive.

These are some of the basic operations. In any specific application there may well, of course, be others (for example, transaction logging, audit trails, etc). We will now examine the major file organisation and processing methods in the light of the characteristics and procedures described above.

Sequential File Organisation

In a sequential file all of the records are stored in a specified, known order. The order, or sequence is determined by the value of some particular field or data item within each record, usually referred to as the *key field*. For example, in the personnel file the records could well be stored in ascending works number sequence. If the works number included, say, a two character department code followed by an employee number, then the records might be sequenced by department, and by employee number within department. This is illustrated in Figure 5.2.

In a customer file, the sequencing key might be the customer code, whilst in the stock file it could be the product code. Note that the field used for sequencing need not be just numeric. Typically it includes alphabetic characters as well (i.e. it can be composed of any alphanumeric characters — usually any of the characters from the standard character sets mentioned in Chapter 3, though often excluding most of the special characters). However, it is usually necessary for such key values to be *unique*. As all retrieval and processing is based on the key value, non-unique keys would give retrieval problems. Thus, in our example of the personnel file the key field would be the assigned works number (which is presumably unique), rather than using, say, the employee name (which, for a large file, almost certainly won't be unique).

```
PERSONNEL
    FILE

Record 1
    works no.      00352
    name           Smith J. R.
    address        13 Acacia Drive
                   Little Bishwick

Record 2
    works no.      00473
    name           Jones F.J.
    . .  . .        . .  . .
                    . .  . .

Record 3
    works no.      00487
    name           Allen A.B.
    . .  . .        . .  . .
                    . .  . .

. . .  . . .
    . .  . .        . .  . .
    . .  . .        . .  . .
    . .  . .        . .  . .
                    . .  . .

End of file
```

Figure 5.2 *An Example of Sequential File Organisation: Personnel File with Records in Ascending Works Number Sequence*

A sequential file ordered by one field will in general not be in any particular order if any other field is used. This may pose major constraints on the way the file is used. For example, if the personnel file was sequenced by works number, but for company records a list of employees in alphabetical sequence was required, then this could only be achieved by sorting all the records in the personnel file such that they were now in sequence by employee name — in essence, creating a new file with exactly the same contents as the existing personnel file, but with the records ordered in a different sequence. Clearly this would appear to be a wasteful exercise (in both processing time and storage capacity needed) if it had to be done at all frequently.

Sequential File Processing

Retrieval of a specific record from a sequential file involves starting at the beginning of the file, reading each record in sequence from secondary memory into primary memory and then comparing its key field with the key field of the record it is desired to retrieve until a match is found; or until the value of the key field on the file is bigger than that of the desired record, in which case the record doesn't exist on the file (assuming the file is in ascending sequence). If there is only one record to be processed, then this is a very time-consuming and relatively expensive operation in computer terms. Because of this, transactions to be processed against a particular master file are usually accumulated in a batch over some period. The length of period may be determined by the operational necessity of the business — such as producing a payroll each Friday, or having an up-to-date stock file available at the beginning of each working day, or simply by waiting until a large enough volume of transactions had been accumulated to make a processing run economic.

As we saw earlier, initially the only form of secondary memory available for computer systems was magnetic tape, on which all files are sequential simply because of the nature of the media itself. Thus the original method of processing in commercial computer installations was almost invariably some form of batch processing. It is still true today that a majority of business data processing involves sequential files (typically stored on disk) in a batch-processing environment. Batching together a number of transactions to be processed against a specific file means that the entire batch can be processed in a single pass through the file, assuming that the batch of transactions has first been sorted on the same key sequence as the master file. The basic logic of a sequential processing run is shown in Figures 5.3a and 5.3b.

The run begins with an *old master* file which is to be updated (this file is the file that results from the previous processing run).

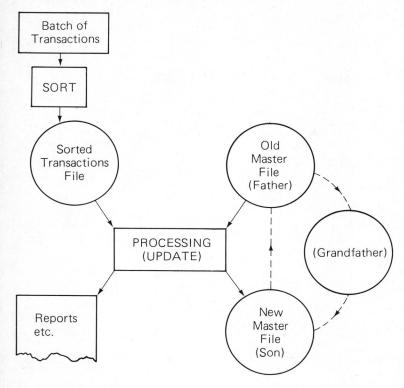

Figure 5.3a *Sequential File Processing*

The batch of *transactions* to be processed is *sorted* into the same *key sequence* as the master file to form a transactions file, in which the records are ordered in the same sequence as the master file. This transaction file is then processed against the old master file (by comparing key field values until a match is found) to create a *new master* file and produce any reports that are required. As a result of the processing, the fields in the records processed on the master file have been updated. For example, in the personnel file the fields containing gross pay to date and tax paid to date will have been updated as a result of a payroll run. All records are read from the old master file and copied to the new master, even if they are not updated. Thus problems of back-up and

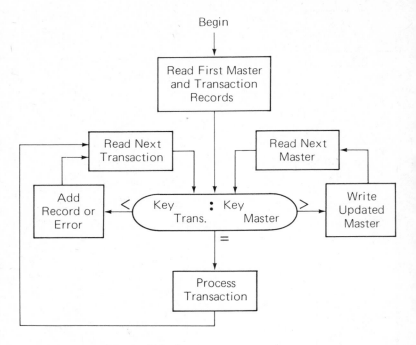

Figure 5.3b *Basic Logic of Sequential Processing*

archiving are simply organised with a magnetic tape-based sequential processing system, since old master files (usually referred to as 'father', 'grandfather' etc, with the new master being 'son') can be retained, together with the most recent transactions file, so that a particular set of processing runs can be repeated if necessary (i.e. the clock can be put back).

  In terms of economic and efficiency characteristics, as might be expected, sequential files have both major disadvantages and advantages. On average the response time is slow where sequential processing is involved, since on average the required record will be half-way down the tape, and all previous records must be read and compared before it is retrieved. The file can only be sequenced on a single key, so that in cases where the file has to be shared by several applications, either they all have to use the same key, or duplicate copies of the file are kept, sorted by another key. For a volatile file, creation and deletion of records is easily accomplished as a part of a normal processing run. If the activity rate is high (i.e. a typical transaction file contains a high proportion of the total records on the file) then sequential processing is very efficient, since most records need processing, and all of the records in the file are accessed in a single run. However, conversely, if the activity rate is low, then sequential processing is relatively inefficient, since all records are accessed, but only a few require processing.

Direct Access Organisation

We have already seen the advantages and disadvantages of using disks as secondary memory devices, and we now turn to direct access file organisation. In recent years there has been a very rapid expansion in on-line interactive systems in all sorts of applications, where fast response to file processing requests is an important design criteria. Some form of direct access file organisation is essential in order to meet such demands. This requires that some method be found for directly computing the physical location of a particular record, given its logical position within a file, as indicated by some specific key value. Given a physical address on a disk surface, in terms of the drive, cylinder and surface, then the disk drive incorporates the appropriate control logic directly to position the read-write arm over that particular cylinder, and to activate the head appropriate to the particular surface. The problem in the design of a direct access file is to specify precisely how this transformation from a record key value to a physical disk address is to be accomplished. There are basically two methods by which this can be done; either *random* or *index-sequential* file organisation. Both enable a particular record in a file to be retrieved directly, but have different operational characteristics and economics. Both enable the 'next' logical record to be read directly without having physically to read the (sequentially) following records.

'Random' File Organisation

This method of file organisation uses what is known as an indirect method of addressing the physical records on disk. Some simple procedure is used to transform the key field value of the record into a physical address. An area of the disk is set aside for storing the file, and then some addressing procedure is devised which will evenly distribute the records over the area of disk. To retrieve a record under this method, the value of the key field for the required record is used by the transformation procedure to calculate the disk address; this is then used to retrieve the record directly. It may happen that two key values give the same disk address (this is known as a *synonym*). As two records cannot exist in the same physical location one has to reside in a designated overflow area. Thus when a record is directly retrieved using this method, the value of the key must be checked and if incorrect then the overflow area is searched (usually sequentially). This procedure is known as 'random' file organisation since the transformation procedures are usually chosen such that the logical records are randomly (evenly) distributed over the disk addresses, and such that the expected number of synonyms (i.e. overflow situations) is also known. This randomising is often referred to as a *hashing algorithm*.

With random file organisation and processing, records can be retrieved in any sequence. Indeed, even if they are to be retrieved in logical sequence (i.e. key value sequence), they still must be retrieved individually since they are not stored in any physical sequence on disk, and there is no way of going directly from one record to the next in key value sequence. Thus, batching or sorting transactions to be processed against a random file does not provide any advantage. The elapsed time and cost of retrieving a record from a random file is roughly constant — response time can be very fast. However, if activity rates are high, processing may be relatively very expensive compared with a sequential file, since processing cost per transaction on a sequential file decreases as the activity rate increases. If the file is highly volatile, then it is likely that a large number of synonyms will be generated. Also, deletions may well leave records in the overflow area when their synonym has been deleted. Thus a volatile, randomly organised file will need very regular reorganisations (usually referred to as 'garbage collection'). Integration of several applications requiring multiple file accesses for one transaction is straightforward and easily accomplished with a random file organisation, since for each transaction the disk addresses in the relevant records and files which it impacts on can be computed and multiple files immediately updated.

However, the overwhelming advantage of a randomly organised file is that it can provide very fast response times (within relatively narrow margins) to file accesses on very large files. It is typically used where these are on-line enquiries onto a large file, where each enquiry initiates relatively little processing, but speed of response is vital. Seat availability on particular airline flights would be examples where such organisation may well be used.

Index Sequential Access Method (ISAM)

Thus so far we have two forms of file organisation which are best suited to very different types of processing and application needs. Sequential organisation and processing gives the lowest cost per transaction for files with high activity rates and is simple to implement — however, the application must not have a critical response time factor. Random organisation and processing is ideal where fast response is essential, and especially where activity rates are low. However, for a large number of applications, some method combining the characteristics of these two methods is required. The problem with the two methods as they stand is that it is an either-or choice — a file organised randomly cannot be processed sequentially.

The ISAM method (sometimes referred to as KSAM: Keyed Sequential Access Method) is an attempt to provide the best of both of the previous systems. ISAM allows for high-activity economical sequential processing, as well as direct access, rapid response processing when required. The way it does this is to provide a set of *pointers* in a *chain* to indicate the next logical record in sequence (usually both next in ascending and descending sequence) relative to some specific key value. These pointers are stored as a part of the record, so that retrieval of a particular record also produces the disk addresses of the records logically adjacent to it in sequence. It is possible to store more than one set of pointers with each record (though this naturally increases the storage used), so that it is possible to access records in the file according to several different keys. Usually, in a situation where multiple keys are used, one key is designated as the primary key, and the file is often physically organised as a sequential file on this primary key. This is illustrated in Figure 5.4.

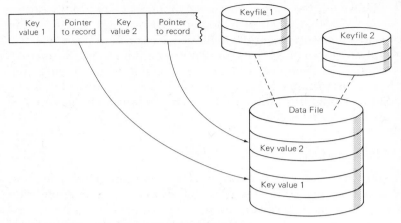

Figure 5.4 *Index Sequential Access Method*

Thus sequential processing is possible on any of the keys by simply retrieving the first record in the sequence for that key, and then following the chain of pointers, retrieving each record in sequence. If this sequential processing is on the primary key, it will be relatively efficient, since the file will be physically in this sequence. In order to be able to undertake direct access processing on such a file we have to have some method of entering a chain close to where the desired record is located, rather than starting at the beginning and working sequentially through it. The way this is done is to form one or more hierarchical indexes to the file. Each index is relatively short and is accessed sequentially. In practice they are usually kept in primary memory to facilitate rapid access. The highest level index provides a broad categorisation of a file, the computer sequentially searches this index to find which category the desired record is in and this in turn points to a second index, divided into further sub-categories which the computer also searches sequentially. The computer may search one or two more indexes of this type, and is then pointed to a record in the file which is close (within a known number of records) to the one being searched for. This record is retrieved and compared with the desired key value — if the desired key value is higher than the value of the one retrieved — then the computer follows the pointer chain sequentially 'upwards' through the file until it finds the one it wants. If the desired value is lower, then it similarly follows the chain 'downwards'.

As might be expected, the economics and characteristics of ISAM files lie between those of sequential and random files. Compared with either sequential or random files solely carrying out sequential or direct processing, ISAM processing will appear relatively inefficient and more costly. However, the user is in essence paying for the ability to carry out both types of processing on the same file if necessary. Naturally, as far as possible it is a good thing if the user can ensure that the high activity transactions are processed sequentially, and the low activity ones directly. ISAM will not provide such a fast response time as a randomly organised file, as it must search one or more indices, and perhaps retrieve several records before finding the desired record. Because of the indexing and the ability to use multiple keys, ISAM files are relatively flexible — for example, often the primary key is used for sequential processing, and one or more of the secondary keys used for direct processing. The user pays for the flexibility through lower efficiency. Because ISAM files are totally dependent upon the indexes and pointer chains, they are potentially more susceptible to software and design problems, and secure back-up and archiving requires much more care. Through the ability to cope with multiple keys, ISAM-organised files also facilitate the sharing of such files between different applications. Again this tends in practice to make them potentially more unreliable. Nevertheless, it is probably true to say that by far the majority of on-line processing is undertaken using some form of ISAM file organisation.

Database Processing

However, all of the file organisation and processing procedures that we have discussed have left the choice of method and implementation to the designer of the particular application system. Thus in many installations, the specification of file structures and organisation was locked into a specific application program, making modification, or use by anyone else very difficult. The same data was often stored more than once, giving consistency problems. In addition it was very difficult to specify or implement common standards for file structures, data accuracy, and so on; and there was a danger of each application re-inventing the wheel in the form of developing particular software to perform standard functions involved in file mainten-ance — creation, deletion, and updating of records, and back-up and recovery procedures. Allied with a realisation of the grow-ing importance of data as a corporate resource, these problems and issues lead to the acceptance of the concept of a *database*, and the development and implementation of specific *database management systems (DBMS)*.

The database concept breaks the direct link between the data files and the applications programs, by inserting some software, a DBMS, between the two. All requests for file access from an application program are now made to the DBMS. The DBMS has incorporated tools enabling it to meet these requests, whilst also undertaking all of the usual file management and maintenance tasks. Schematically Figure 5.5 illustrates the differences between a more traditional file organisation and the use of a DBMS.

As a part of, or closely associated with the DBMS will be several other pieces of software. A Data Dictionary will maintain a record of each data-item held in the database and its character-istics (i.e. type and length); thus the DBMS 'knows' about all the data-items in the database. The DBMS itself will know the logical structure of the records in each of the files within it, and the relationships between them. It would reduce, if not totally re-move, duplication of the data in different files, and enable data to be much more easily shared where necessary. Requests for access required (e.g. either read-only or update) can be control-led. Back-up and archiving can be undertaken automatically at specified times, and transaction logs and audit trails generated as a part of normal processing. For input of data, a screen-handl-ing utility is typically provided which will generate screen formats, and undertake some verification of the data provided, as well as interfacing with the DBMS. Usually, the utility can be used directly by the users to create and modify their own screen formats. On the output side, an English-like query language allows for *ad hoc* enquiries against the database, handling for example, queries such as:

FIND CUSTOMERS FOR REGION = NW AND

BALANCE-OUTSTANDING > 10000

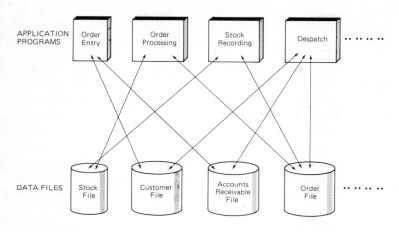

Figure 5.5a *Relationship between Application Programs and Files using a Traditional dp Approach*

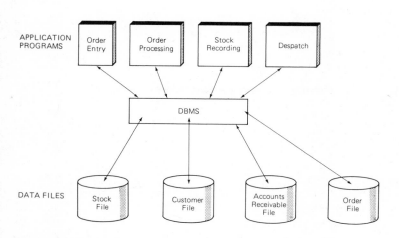

Figure 5.5b *Relationship between Application Programs and Files using a Database Approach*

On the other hand a report generator would facilitate the design and production of more regular and formal reports. The relationship between a DBMS and these utility programs is shown in Figure 5.6.

A DBMS and its associated utility programs are large, complex pieces of software. From the schematic diagram shown in Figure 5.6, it should be apparent that the DBMS will be extremely heavily used, and from the user's point of view will have a major impact on overall system performance. DBMSs do consume large amounts of computing resources — both primary and secondary memory, and processor time — and in general will impose some additional system overhead as compared with a more traditional approach. However, these additional costs are generally considered to be more than outweighed by the benefits of DBMSs as

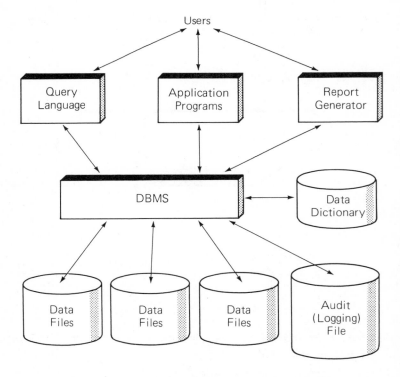

Figure 5.6 *Functional Components of a Database System*

outlined above, and perhaps above all, by the flexibility and ease with which new systems can be developed.

DBMS were originally developed for large mainframe computers during the early and mid-1970s, both by the computer manufacturers and by third-party software houses. Today DBMS are available for all of the major large-scale computer systems; for example from the computer manufacturers, IMS from IBM, and DMS from Univac; and from software houses, ADABAS from Software AG, IDMS from Cullinane, DBMS-2000 from MRI Systems, and the most successful system of all, TOTAL from Cincom. Furthermore, the scale of systems on which DBMSs are available now includes most minicomputers; and indeed, one of the major reasons for the success of Hewlett Packard in growing from an instrumentation and measurement company into a more broadly based computer manufacturer was the availability of its DBMS, Image, as a part of the standard system software for its range of minicomputers. Few organisations would consider developing new dp applications without using some form of DBMS. The problem lies with the management, maintenance and possible conversion of the existing, more traditionally organised application programs and their associated data files. This is a topic which we will return to discuss in Chapter 14 in the context of the problems in software development.

This brings us to the end of our discussion on computing and the components of data processing. As all this might seem quite remote from the prospect of say, home computers or the electronic village, let us now look at the extraordinary developments in the underlying technology that are bringing the computer into contact with everything and everybody.

Part III

THE ADVENT OF THE MICROCHIP

In the previous part we reviewed the fundamentals of computer systems — both the characteristics of the hardware components, as well as the nature of the software and its critical role in providing a relevant application system to the user. The prime reason for the rapid development of information technology has been the phenomenal technical progress seen in microelectronics over the past 25 years or so. One of the most significant results of this progress in microelectronics is that the computer has become a low-cost, reliable component in an immense range of products — in many instances a component that the user is totally unaware of. The computer has become an all-pervasive bearer of intelligence to an ever increasing variety of products. It is this low-cost, pervasive computer as a component which is providing much of the driving force in the development of the information technology industry.

Any products or processes involving measurement or control, together with those doing signal or data processing at some point are now potential candidates for microelectronics based design. Microelectronics can cram the processing power of a major computer into components so small and at a cost so low as to have been incomprehensible a decade ago. (Indeed if Neil Armstrong had found a 1981 wristwatch calculator on the moon he would have been staggered by the capability contained in such a small space.) Chips, the product of microelectronics, have been likened to the crude oil of the 1980s as the critical national resource, and it has been suggested that micro-electronics lies a close second to the wheel in terms of social, commercial and industrial significance. Few spheres of human activity and few sectors of industry are likely to remain untouched by the application of this technology. So in this part we will try to explain 'what is a microchip' by discussing:

the major technical developments in microelectronics;

the basic economic characteristics of the production process;

the major product categories;

the development and exploitation of the market for the products of the industry;

and the present and prospective applications of this technology.

6

The Technological Base

The Development of Microelectronics

Valves

What we term microelectronics is the current state of a continuing evolution in the development and manufacture of electronic
components. *Passive* components, such as resistors, capacitors
and coils provide electrical friction, storage and inertia, while the
active components have two functions — amplification and
switching. Most signals in the real world vary continuously in
amplitude and frequency. The obvious example is sound: it is
either loud or quiet, high pitched or low. If we wish to capture
this signal electronically, then the easiest way to do this is to take
an electrical *analogue,* i.e. a voltage or a current that mirrors the
signal exactly. This is exactly what a microphone does. Similarly,
a loudspeaker reverses this process.

Most signals, once in electrical analogue form, need amplify-
ing. How accurately the amplified signal represents the original
will depend on the linearity of the amplifier. The other role in the
active component is switching. The key factors here are speed
and 'cleanliness'. Speed will depend mainly on the electrical
inertia — it's like the difference between closing a barn door and
a cupboard door. Cleanliness is the ability to switch accurately
from one state to another without overshooting.

The basic building block of all electronic digital devices is an electronic switch or 'gate', which allows the passage or otherwise of an electrical pulse, depending upon the instruction of some control unit. Originally, all electronic devices were assembled using *valves* (vacuum tubes) to form these electronic switches, the characteristics of which were:

> individually they were relatively large components;
>
> they required large amounts of power for their operation;
>
> they generated considerable amounts of heat;
>
> they were very unreliable in operation.

Thus the early computers and other items of electronic equipment were extremely bulky, needed carefully controlled physical environments, and were very expensive to purchase (since the computer or other electronic device had to be assembled by hand from tens of thousands of individual components) — yet the average time between breakdowns for such early computers was usually measured in minutes!

Transistors

The successor to the valve in forming electronic switches was the discrete *transistor*, originally invented in 1946, but because of technical difficulties not in mass-production until the mid-1950s. Initially the first transistors were made individually by hand, and were much more expensive than the corresponding valve equivalents. The relative advantages over the then current valve technology were:

> much smaller physical volume;
>
> much lower power requirements (enabling batteries to be used as the power source);
>
> and much greater robustness and reliability.

These led to their initial use in a few products where these characteristics were sufficiently important to overcome the higher cost penalty — for example, domestically in hearing aids and for the military in missile guidance systems. Subsequent reductions in cost then led to the boom in products such as transistor radios, where portability, small volume and reliability gave the transistor-based product an unassailable advantage.

Development thereafter followed a familiar pattern and one that has been repeated on several occasions since; increased volume of production led to lower average costs, which were then passed on to the consumer, thus expanding the market, lowering costs still further, generating further expansion of the market, and so on. Each step in technology has ultimately lowered the cost per circuit to the end-user, and has consequently ruthlessly wiped out its predecessor. For example, the response of the valve manufacturers to the mass-produced transistor was to try to automate the production of valves; the result was that, of the top 10 valve manufacturers in the 1950s, only one survived to be a major transistor manufacturer of the 1960s.

Integrated Circuits

The next major technological jump came with the development of the planar transistor in 1957 (i.e. a transistor manufactured as an integral part of the flat surface of a piece of material), about a ¼ inch square, and 0.015 of an inch thick. Since this date there has been continuous progress in the search to squeeze ever more and more circuits into a given surface area. Hence the overall generic term for what we are discussing is literally *microelectronics*, the design and manufacture of ever smaller and smaller electronic devices, each incorporating many electronic circuits. The basic unit is the *'chip'*, typically a rectangle of a *semiconducting* material (usually silicon, hence silicon chips), in which is manufactured many tens of thousands of electronic circuits. The industry which manufactures these devices is similarly known as the *semiconductor industry*.

Any device consists of a set of circuits to perform a certain function which, when combined with other devices, forms the final electronic product (which itself is normally only a component of the final consumer product) — for example, the memory and arithmetic components of an electronic calculator; the sensing and control circuits for an automatic washing machine; the signal decoding circuits for a hi-fi tuner, and so on. The trade name for chips incorporating more than one circuit is ICs (*integrated circuits*), and the scale of integration is indicated by the use of the following mnemonics:

SSI   Small Scale Integration (1–20 gates)

MSI   Medium Scale Integration (20–100 gates or 1K-bits memory)

LSI   Large Scale Integration (100–10000 gates or 16K memory)

VLSI Very LSI!

(The majority of computer-related mnemonics are this simple!) The relationship between these definitions is illustrated in Figure 6.1.

Obviously these terms are relative — what is LSI today may seem only MSI tomorrow — but there is general consensus that the industry is presently moving into an era of VLSI, enabling all of the circuits of a large-scale computer system to be manufactured in a single chip. The difference between integrated circuits and what went before is not one of kind, but one of degree. Previously each transistor, diode or whatever had to be manufactured separately and then wired together on a board to form the required circuits. On a chip tens of thousands of circuits are manufactured at the same time, together with all their circuit interconnections. Greater and greater degrees of integration mean more and more circuits mass-produced jointly and simultaneously.

So, what are chips? How are they made? Exactly what functions can they perform? The next two sections of this chapter will examine these questions.

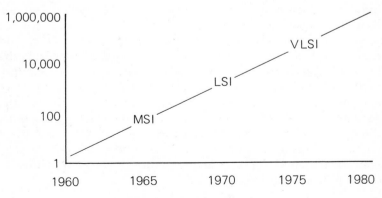

Figure 6.1 *Packing Density*

Manufacturing Process

Chips are manufactured using as a base a semiconducting material (usually silicon), which can be impregnated with impurities such that it will function as an electronic switch, being moved electronically from one state to another in a few millionths of a second, either allowing an electrical pulse to pass or not, as the case may be. Basically this is achieved in two steps. First, the silicon is 'doped' by different chemical compounds. The two common variants are what is known as 'n-type', in which typically arsenic or phosphorus is introduced to bond with the silicon atoms to leave a free electron (which can be induced to move); and 'p-type', in which, typically, boron is introduced to bond with the silicon atoms leaving them one electron short (i.e. leaving a 'hole' into which other free electrons may be induced to move). The circuits to be formed on the silicon (transistors, resistors, diodes, capacitors, etc.) are manufactured as a series of n- and p-type regions. The second part of the production process is then to provide the necessary electrical insulation and connections between these circuits.

This is a description of what is known as a *Field Effect Transistor* (FET), which is manufactured using a metal-oxide on silicon process (referred to as MOS). The other major approach to circuit design is known as *bipolar*. The basic difference between these two approaches is in power requirements and speed of switching — MOS has lower power requirements but switches more slowly. The latest development in FET technology, CMOS, improves on switching speed whilst maintaining lower power requirements.

In the manufacturing process, pure crystals are grown from a furnace containing molten silicon (as long salami-like cylinders), sliced into extremely thin *wafers*, and then polished. The diameter of these silicon wafers on which the circuits are to be manufactured is largely a technical constant (at about 3 inches, though some manufacturers are reportedly moving to 4-inch wafers and larger). The circuits to be manufactured on the chip are first drawn out on paper to a very large scale (about 250 times larger than true size, a process done by hand originally, but now carried out by computer), and then photographed and reduced in scale. The various patterns of layers of circuits making up each chip are then repeated between 200 and 400 times for each chip on the wafer, to form a set of production masks for the wafer as a whole. These *photolithographic* masks, reduced in scale by many orders of magnitude, are used in the chemical etching process by which the circuits are manufactured in the silicon. A series of 20 or 30 masks and sequential etching processes may be required in the manufacture of a complex chip.

At the start of the etching process (which is illustrated in Figure 6.2) the surface of the wafer is given an oxide coating to provide a basic insulation layer (step 1 in Figure 6.2). The regions to be doped in the first layer of the circuit are defined on the surface of the wafer by the first mask (step 2 in Figure 6.2). The surface is given a photosensitive coating and the mask laid on it — the regions to be doped are masked; those to remain undoped are not. The surface with the mask on it is exposed to light — the photosensitive material in the unmasked regions becomes 'fixed' (i.e. it is permanently bonded to the surface of the chip), while the unmasked surface remains soluble (step 3). The wafer is then dipped in acid which etches away the soluble, unmasked areas, creating a set of 'windows' to be doped (step 4). Next, the wafer is placed in a furnace in which it is exposed to an atmosphere containing the required dopant, and the dopant diffuses through the 'windows' into those areas of the wafer (step 5). This sequence is then repeated many times to build up a layer of circuits on each chip on the wafer. The final mask provides the windows to be etched to provide the contact points for the metal interconnections between the circuits on each chip (step 8).

The manufacture of the chips on the silicon wafer is only the first in a long sequence of processes that must be undertaken before a product useful to the customer arises. Something of the order of up to 400 chips may be manufactured on a single wafer, and each wafer must be broken up into its constituent chips before each chip can be mounted and fitted with connectors. At this stage the chip can be relatively easily handled, connected up to power supplies and other chips, and assembled into the final component. Originally this was usually undertaken in the Far East (typically in Singapore, Korea, Taiwan and the Philippines), as it is basically a manual operation. However, several companies have now automated major portions of the chip testing and mounting process, especially for chips used in standard components (e.g. computer memories), and this is a trend that will undoubtedly continue.

1. Silicon dioxide layer (insulator)

Silicon wafer

2. Photosensitive layer

Ultraviolet radiation

Mask
Pattern

3. Hardened layer

Photosensitive layer is etched
away in non-hardened areas

4. Doping and further oxidisation

p-type doped areas

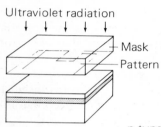

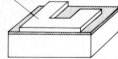

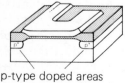

5. Unwanted photo-resist
is etched away

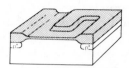

6. Second masking and etching step to
deposit first polysilicon layer
First polysilicon layer

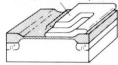

7. Second masking step, and doping

Second polysilicon layer

n-type doped area
Insulated oxide area

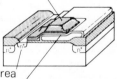

8. 'Contact' windows are opened up
for the metal (aluminium) contacts

Contact window

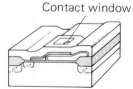

Aluminium

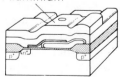

Figure 6.2 *Stages in the Chip Manufacturing Process*

Before the 'breaking-out' and mounting process, the chips are tested — in a process by now almost totally automated so that the costs of mounting and subsequent handling are borne only by the good chips. For the really complex processor chips, however, full 100% testing is impossible, because of the enormous number of possible combinations of instructions and data that might occur. Some manufacturers, particularly the Japanese, are charging premium prices for more highly tested chips, and for a price will supply guaranteed 100% reliable batches. (For example, this is alleged to be one of the reasons for the apparent greater reliability of Japanese television sets a few years ago.)

Packing Density

The major design and technical challenge facing the semiconductor companies is either to reduce the physical dimensions of the chip (yet still accommodating the same number of circuits); or to increase the number of circuits (for the same size chip). Both of these two cases are but alternative manifestations of the same process, namely, increasing the *packing density* on the chip (i.e. more circuits in a given space). The first chip, containing a single transistor, was produced in 1957. In 1964, it was suggested that the future packing density would continue roughly to double every year (as illustrated in Figure 6.1). This prediction certainly held true up to about 1975, but there is some evidence that since this date there has been a slow-down in the rate of innovation, and in 1979 it was suggested that the rate of progress was now equivalent to a doubling in density about every two years. (Some of the possible reasons for this slow-down are discussed below.)

Some deviation from exponential growth is eventually inevitable with current technology, but technology in general is still some way from the limits imposed by the fundamental laws of physics. The packing density increases partly through improvements in the design of the circuit on the wafer, but largely through improvements in manufacturing technique, the ability to etch ever finer and finer detail into the silicon. At present photolithographic methods are used, but the limits of this technology are rapidly approaching, and methods using beams of electrons to etch the circuits directly are currently being implemented (the problem being that the wavelength of visible light is too large to draw lines only one or two atoms wide!).

Present technology enables manufacturers to etch transistors in the silicon about 2 to 4 microns long (a micron is one thousandth of a millimetre), and over the next few years this will be progressively scaled down to about 1 micron. The width of the metal providing the electrical contact between the transistors on the chip is currently about 5 microns — similarly it should be possible to get this down perhaps to 2 microns. Beyond this point there are problems in ensuring electrical continuity in the metal, since gaps may appear between the individual atoms of the metal forming the conductor! Improvements such as these are perfectly feasible in the next few years using presently known techniques, and would enable manufacturers to pack between four and ten times as many components on a chip as at present.

Production Control

The major problem facing the semiconductor manufacturers is to increase production *yield*, by reducing the number of defects introduced during the manufacturing process. The major benefits in increasing production yield come from reducing the incidence of defects in the final chip. The minutest particle of dust is sufficient to wreck a chip; and thus with the ever-increasing need to achieve mass production has come detailed attention to the production environment. This environment needs to be many thousand times cleaner than the best hospital operating theatre, and is, therefore, very expensive. In the early years of the development of a particular chip design, it is not uncommon for the defect rate from an integrated circuit production line to be as high as 90% (and instances of 100% defect rates have been known to occur). Today it is normally at least half this rate for established chips, and any manufacturer who gets down to 20% will be very pleased. It is still true that the defect rates for each new chip follow a curve, starting high and declining through its life — the name of the manufacturing game is to reduce the initial height of this curve.

For example, in 1980, Motorola was prototyping its latest 16-bit microprocessor chip (the MC68000), probably at the time the most complex chip in the world. It contains about 68000 transistors on a piece of silicon measuring 246 by 281 mils (a mil is a thousandth of an inch!). Of the hundreds of chips on each wafer that Motorola was producing, probably only two or three were good ones, and the remaining few hundred were scrap. It was early 1981 before it achieved mass production of the order of the thousands of good chips per month necessary to make the operation profitable. As an illustration of the continuing pace of development in this area, by early 1982 Hewlett Packard had announced a chip containing the equivalent of 450,000 transistors.

Yield is also a function of the complexity of the logic design on the chip and of the packing density. Thus for new, very complex and dense VLSI processor chips, yields may only be of the order of 10–15 good chips per wafer (i.e. a 90% failure rate) which is reflected in prices of $50 or more each. The other problem for the manufacturer is that the yield is unpredictably variable — good one day, terrible the next, and occasionally terrible for weeks! Failure to achieve reasonable yields fairly quickly can obviously have serious financial consequences. Even the giants are not immune: IBM encountered considerable difficulty in obtaining adequate yields of its new 64K memory chip in 1980, and was forced into having to buy large quantities of the then standard 16K memory chips from the semiconductor companies.

So what are the electronic circuits that are manufactured into a chip?

Semiconductor Products

The products of the semiconductor industry can be broken down into chips which perform one of four major types of function, namely:

> non-digital logic components, comprising about 25% of the market for semiconductor products;
>
> digital logic components, comprising about a further 25%;
>
> general-purpose microprocessors, comprising about 10%;
>
> digital memory components, comprising the remaining 40% of the market.

We will describe the characteristics and development of each of these categories in turn.

Non-Digital Logic

By 'non-digital' we mean circuits that do not handle binary-coded information, but rather those that handle information by representing it as a varying electrical voltage, i.e. analogue circuits. At one time almost all electronic components were made up using analogue components. But, slowly at first, and then with increasing rapidity in recent years, binary-coded digital representation has taken over, particularly to improve accuracy in processing and transmission. However, there are still many electronic components operating in analogue mode, for example, in situations where it is necessary to control electrical currents. Increased integration of circuits on a chip has progressed in this application area as much as in any other, typically giving the manufacturer of products using such chips the ability to use fewer discrete components overall, represented by replacing many discrete transistors, diodes etc. with a few integrated circuits.

As an example of the changes that increased integration can bring, Figure 6.3 illustrates how the component count has changed in recent years within a typical product — a 20-inch colour television set — which incorporates non-digital circuits. As can be seen, the number of individual discrete components has fallen from a total of 230 in 1970 to less than 141 in 1978, whilst the number of integrated circuits has increased by only 7. This brings major benefits to the manufacturer in both assembly time and in the resulting relatively lower production costs, together with the improvements in the overall reliability of the final product.

	1970	*1978*
Integrated Circuits	2	9
Transistors (Discrete)	65	34
Diodes (Discrete)	65	38
Miscellaneous Circuits	100	50—70
Power Consumption	155 W	85 W

Figure 6.3 *Components in a 20-inch Colour TV Set*

Digital Logic

By digital logic we mean those components that process, transmit or otherwise act upon instructions presented in binary-coded digital form. Digital logic devices are special-purpose processors, designed to optimise their performance in one specific application. Hence they are also known as *custom logic* or special logic chips, to distinguish them from general-purpose microprocessors. They find typical application in products to fulfil a control function — for example, in missile guidance systems, signal processing, process control, and telecommunications equipment. The functions incorporated in such a chip are fixed at the time of manufacture: very specific processing capabilities are actually incorporated into the chip at the time of its manufacture, and cannot subsequently be modified (i.e. the program that it contains is fixed and it cannot be reprogrammed). Thus, the customer has to be sure that the design is correct, and since the chip has only very specific application, it is usually economically worthwhile to produce such chips only when the volume required is very large. As the user is in effect having a chip made to his specific custom design, the user may well have to bear all of the design costs, which for a complex chip, can be extremely expensive.

The prime benefit to be gained from using such components rather than the cheaper general-purpose microprocessors arises from their purpose-built design. The most important benefit is usually speed of operation (although low power consumption or a wider operating environment are of critical importance in some applications). For example, a purpose-built processor tailored to a specific processing application might operate ten times faster than would a general-purpose machine programmed for the same application. However, the major penalty that has to be paid for such performance is typically design lead-time. The whole point about devices such as these is that they are special-purpose, explicitly designed with a particular application in mind, so that the design lead-time is usually a minimum of one year, and more typically two. In contrast it is perfectly possible to design an application using a standard off-the-shelf microprocessor in a fraction of this time, certainly well within a year.

A recent technical development which may well speed up this application development lead-time is the availability of what are known as *uncommitted logic arrays* (often called *gate arrays*). These are digital circuits in which the precise logic to be followed is not determined at the time of manufacture (it is uncommitted), but can be permanently fixed subsequently by the user (or by the chip manufacturer on his behalf). However, once the particular logic pattern has been fixed in such a device it is permanent, and it cannot be reprogrammed as a general-purpose microprocessor could be.

The benefits are considerable — the manufacturer can make one uncommitted logic array chip which could cope with the processing logic formerly incorporated in several specific individual chips, thus increasing his volume for a single device and lowering his costs. From the point of view of the user, the point of time in the development cycle at which the logic design has to be fixed for manufacture is put back considerably, since the user is no longer having a specialist chip made, but is having a more general-purpose mass-produced chip configured to his specific requirement. Also, as the user is now making use of a more widely applicable chip, rather than a custom design, his design costs will probably be much lower. For example, both the Sinclair ZX81 and the BBC microcomputer incorporate components of this type, manufactured by Ferranti (who were one of the pioneers in this technology).

General-Purpose Microprocessors

These chips are basically general-purpose versions of the digital logic chips described in the previous paragraphs. They are *general-purpose* in that the function that they perform is fixed by software — by a program characterising the user's application — be it running a digital watch, controlling a washing machine or functioning as a word-processor. From the user's point of view it is this *user programmability* that is attractive in using such chips. He can buy a cheap, standard off-the-shelf component, and directly program his application requirements.

Further, the manufacturers of standard microprocessors typically supply a complete 'family' of other chips to handle a wide range of functions necessary to make use of the microprocessor in practice, for example, various input/output controller chips to interface the processor to a variety of instruments, sensors and actuators. Many provide the chips necessary to form a complete microcomputer system. The availability of these other off-the-shelf chips may have a major impact on the customer's design process, since it is often the interface between the microprocessor and the host product that takes most of the design time and effort. Because of the design cost and lead-time, the trend is increasingly towards using standard microprocessors, rather than custom logic. It is significant that when it came to applying chips in cars (as described in the next chapter), both General Motors and Ford chose to use standard microprocessors rather than design their own custom logic, despite being very high volume customers.

Microprocessors are classified according to their word-length (i.e. the number of adjacent bits that they process in one operation). The original, and still the mass market for microprocessors is for 4-bit processors, which are primarily used in simple control functions. With prices now down to less than $1 each in the USA for very large quantities, they have been turning up in all sorts of products, particularly in toys and games. In 1978 more than 16 million of these were sold, of which the market leader Texas Instruments (TI) alone sold over 9 million, though rock-bottom prices meant rock-bottom margins too. (But even with small margins, selling 9 million of anything isn't bad!) In the case of TI it was its own biggest customer for processor chips, using them in calculators, watches and toys, segments of the market where the margins were almost certainly better. Most 4-bit chips are used as components on their own with very little additional memory or other logic chips associated with them, and they are not intended to be end-user programmable.

The next product up the line, the 8-bit processor is for the semiconductor companies a different commodity. It is a much more complex chip in which design differences between manufacturers become an important factor in marketing success. Also the add-on market for memory and other chips is enormously larger — it is this processor which has led to the boom in personal computing and small business systems. In these sort of systems, the processor chip is usually accompanied by at least three or four memory chips, together with probably at least another half dozen logic chips for controlling particular peripheral devices. Unlike the 4-bit market, the 8-bit market is not dominated by one company and indeed Texas Instruments did not have a product in this market until very recently. In the 8-bit market the leaders are Intel, Fairchild, Motorola, Zilog and Mostek.

The significance of the 8-bit and upwards processors is two-fold: first, in the software design and support required of the semiconductor companies — and second, in the markets towards which these products are aimed. The control functions to which the original 4-bit processors are applied are so simple that they are fixed at the time of chip manufacture for the end-user. With the 8-bit devices this is still feasible, but the prime attraction is that they are end-user programmable in machine-code or a high-level language and thus can be employed in an immensely wide range of tasks. 8 bits is sufficient to store one character (alphabetic, numeric or special), so that 8-bit processors are ideally suited to text processing applications.

At the present time, the semiconductor companies are geared up to mass-produce 16-bit processors on a chip, and are just starting to produce prototype 32-bit processors (i.e. the equivalent of a medium-sized mainframe computer on a chip!) One of the problems that the semiconductor companies face in marketing such chips is that they are now entering a market area previously dominated by the mini- and mainframe computer companies. Software design and support tools are even more important than for the 8-bit processors, and it is significant that Intel has already announced that the new programming language, ADA, will be incorporated into its new 32-bit processor as a system design language. We will consider some of the problems in identifying the market for the 16- and 32-bit processors on a chip in the next chapter.

Digital Memory

The final, and in terms of market size, largest market for the products of the semiconductor companies is in the provision of memory chips — chips which store binary-coded information. It is important in understanding the characteristics of the various products of the semiconductor companies to realise that there is a fundamental difference between the previous three product categories that we have considered and that of memory. In the case of the products of the previous three categories, from the point of view of the producing companies they can be characterised as being constrained by the skills of the chip designer. If you were to examine one of these chips under a microscope, then there would appear to be little pattern to circuits on the chip's surface — for this reason these chips are known as being comprised of *random logic*.

In contrast, if you were to examine a memory chip in similar fashion the pattern of circuits on the chip would be seen to be very regular — hence they are comprised of *regular logic* — and the constraining characteristic from the point of view of the producing semiconductor company is technology. The constraint on the performance of memory chips is the technical ability of the producer to pack more and more of the same type of circuit in a given space. For logic chips the constraint on performance is the ability of the designer of the chip to incorporate more and more sophisticated logic. Thus logic chips sell on processing performance and particular design characteristics, and have more individual markets, whilst memory chips sell on the basis of cost per bit stored and share a generic market.

Memory chips can be divided into two types: *RAMs (Random Access Memory)* are memory chips which can have data or instructions both written into and read from them by the host processor; *ROMs (Read Only Memory)* are chips from which the data and instructions they contain can only be read by the processor (RAMs form by far the largest part of this market).

The primary memory area of a computer is typically made up of RAM chips — it would be better named read/write memory to describe its true nature, as ROM also has random access capability. A further distinction between RAM and ROM memory chips is that the former is typically *dynamic* — it retains its contents only so long as electrical power is applied to the chip — whereas ROM is *static*, retaining its contents even when power is lost. Thus, many computer systems incorporate a battery back-up capability which will prevent a loss of contents of the computer's memory if power is lost (usually for between two and four hours). It is possible to manufacture static RAM, but it is more expensive than the dynamic type, and there are fewer applications where this higher cost can be justified.

So the state of the art in memory chip design is determined by the capacity of a chip, as measured in terms of bits stored per chip (i.e. the results of improvements in packing density). The improvement in bits per chip since 1975 is shown in Figure 6.4. Note that the dates indicated for each chip show the year in which that chip technology started to appear generally in end-

1975	1K Bits
1977	4K Bits
1979	16K Bits
1981	64K Bits

Figure 6.4 *Capacity of Generally Available Memory Chips*

user products (typically primary computer memory). Thus, 64-bit memories began to be used by the computer companies in 1981, though at the same time one semiconductor company at least was prototyping 256K-bit memory chips.

Whatever is stored in a ROM is fixed (usually permanently) at the time of manufacture of the chip. By incorporating appropriate instructions in a ROM, a software house can take a general-purpose microprocessor and turn it into a system for addressing a specific end-user problem. Also the instructions in ROM are in a form which only the processor can understand, and cannot be altered by the user, so that the investment by the software house in that system is protected. Most computer systems include some ROM — it is increasingly being used to store parts of the operating system of the computer which might otherwise be stored on disk; thus giving a boost to performance, since the access time to ROM is much faster than to disk.

If a standard microprocessor is used in a product, then the instructions that it follows will normally be incorporated in ROM. To avoid the uneconomic nature of very small production runs in manufacturing ROMs, in practice what are known as Programmable ROMs (PROMs) are widely used — these are ROMs which are user-programmed, but can only be programmed once (this is done using a machine known as a PROM Blaster!). An even more sophisticated version of this principle is an Erasable PROM (yes, it's known as an EPROM!) — a ROM which is user-programmable, but can be user-erased (by exposure to ultraviolet light) and reprogrammed a limited number of times. Naturally, both PROMs and EPROMs are more expensive than straightforward ROMs. As a general rule, each additional letter in the mnemonic increases the price by an order of magnitude, so that EPROMs are often priced at $40 and above. However, to the customer this appears to be a very small price to pay for greatly increased flexibility in the design process, especially when placed in the context of the final product, which may well have a value a 100 or 1000 times greater.

Other Technologies

Although most of the popular discussion about microchips is in terms of silicon chips — i.e. those using silicon as the semiconducting base on which the circuits are fabricated — it is important to appreciate that other semiconducting material may be used. For example, several companies, and especially the Japanese, are working on manufacturing chips using Gallium Arsenide as the base material. Chips fabricated on this base have lower power requirements and operate at much higher switching speeds than silicon-based circuits, and are thus of great interest to companies attempting to design faster processors. Problems have been encountered in achieving the very high packing densities on Gallium Arsenide that are possible using silicon, but recently these problems appear to have been overcome.

In the slightly longer run, the next generation of technology following on from semiconductors seems likely to be based on the superconducting properties of some materials when they are cooled to extremely low temperatures (to only a few degrees above absolute zero). Originally discovered by Professor Brian Josephson (and hence known as Josephson junctions), circuits using these materials offer switching speeds many orders of magnitude faster than those achieved by silicon circuits, though some developers suggest that the speed of Gallium Arsenide circuits may approach that of Josephson junctions if they too are super-cooled. However, a present complication is that in order to make these circuits operate, they have to be immersed in a bath of liquid helium in order to achieve the desired low temperatures. But several manufacturers, especially IBM, have put a considerable amount of effort into developing this technology. Indeed, it has been suggested that in the next century the natural place for computers to be located will be on satellites in space — it's very cold, there is ample free solar energy for power, and high-speed communications with Earth (and elsewhere!).

The production skills necessary to fabricate chips also made possible the development of two other types of solid-state memory devices — magnetic bubble memories and charge-coupled devices. At one time it was thought that memories manufactured using one or other of these technologies might fill the price/performance gap between RAM memories and disk memories. However, in the event only bubble memories ever appeared as a commercial product. Magnetic bubble memories — as the name suggests — store and recall chains of minute magnetic bubbles in a chip-like device. They have the same advantages as RAM memory: large storage volume in a very small space and low power requirements, but they never become truly price-competitive with RAM memory. However, they have found application in specific instances where their characteristics are especially useful. For example, TI incorporates a bubble memory storage capability of 96K bits in some of their portable terminals, and Rockwell developed a 1M bit unit in a package about a foot square for use in the space shuttle. The American companies have now dropped their development effort in this area, although one or two of the Japanese manufacturers are still continuing.

7

The Semiconductor Industry:
Its Economics and Characteristics

In the previous chapter we examined the basic technology of the semiconductor industry, and reviewed the major types of circuits that can be manufactured in a chip. In this chapter we consider the consequences of the nature of the manufacturing process, in terms of the economics of production faced by the semiconductor companies, and review the resulting significant characteristics of the industry. First we examine the technical and economic consequences of making electronic circuits ever smaller and smaller.

Small is Beautiful

As has been suggested for economics in general, so definitely in the economics of microelectronics, smallness brings beautiful economic benefits! The primary consequences of making microelectronics even more 'micro' — that is, increasing the packing density — are illustrated in Figure 7.1.

There are four primary areas of benefit. First, the circuits manufactured into the chips are concerned with manipulating and storing electrical pulses representing binary digits. If the packing density increases, then the circuits on the chip are closer together, so that the time taken for the electrical pulse to travel from one circuit to another is less, and therefore the component

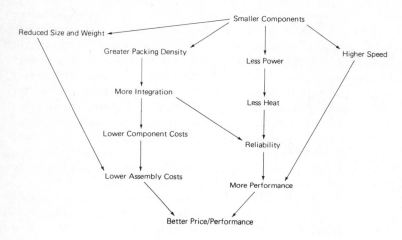

Figure 7.1 *Benefits of Microelectronics*

on the chip will function faster (since the electrical pulse spends less time travelling between circuits). An indication of the increase in speed over the past 20 years is shown in Figure 7.2.

Second, as the circuits on the chip are now smaller and closer together, their power requirements are less, so that, for example,

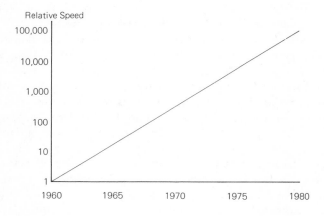

Figure 7.2 *Increase in Speed*

operating from batteries is feasible for a wider and wider range of products. In an electronic component there are no moving parts, so that all of the electrical energy used to switch the circuit from one state to another is given off as heat — lower power requirements mean less heat to be dissipated. This is of direct and indirect benefit: adequate dissipation of heat is a major constraint on the designers of both the chip and the product that it is used in (for example, even today the cpu's of many of the world's most powerful computers are liquid cooled). Also, heat (too much of it) is a significant influence on the reliability of a chip, so that a reduction in the amount of heat given off usually provides an improvement in chip reliability.

Third, an increase in packing density enables a greater number of circuits to be incorporated into a single chip. Operationally, the reliability of chips in general is fairly constant within each of the four major categories, irrespective of the number of circuits on the chip. This is because the chip itself is extremely reliable (especially after a 'burning-in' period of continuous running). The unreliability typically occurs in the connectors between the chip and the board on which it is mounted. These connectors are much more unreliable, being vulnerable to shock, vibration, and corrosion — characteristics independent of the nature of the function of the chip. Thus for the user who requires a fixed number of circuits to perform a given function in a final product, increasing the packing density on the chip means that he will require fewer discrete chips in his final product, and thus proportionally improve the reliability of his product.

The final benefit of increasing the packing density is that the customer gets all this, and at lower and lower prices. More circuits on a given chip means that the cost per circuit falls, since the manufacturing costs of a chip are largely independent of the contents of the chip. Also, for a given function, the user will require fewer discrete chips, so that the cost to the user of providing a given function falls dramatically. The decline in cost is illustrated in Figure 7.3.

In the chip manufacturing process the costs of all 400 or so chips on a wafer are shared as joint costs until the break-out stage. The manufacturing process up to the break-out stage is highly automated (and capital intensive). Establishing a new chip production facility entails heavy investment in specialised equipment (e.g. clean rooms, doping ovens, photolithographic

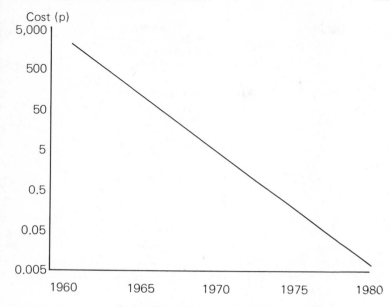

Figure 7.3 *Cost per Component*

facilities), such that roughly 90% of the costs are incurred before a single chip is produced. Hence, it is a mass production business, and most chip manufacturers aim to produce hundreds of thousands of chips per week.

Up until this point, the costs of the multi-stage etching and lithographic processes are borne by all the chips on the wafer, good or bad (hence the importance of production yield mentioned earlier). It is only after break-out that each (good) chip carries its own handling and mounting cost. This is illustrated in Figure 7.4. Recent developments in automated testing, and now progress towards automated mounting of the chip in its carrier and on the board, will serve only to heighten and emphasise this characteristic.

An illustration of how the benefits arising from the economics of chip production are brought to the user is shown in Figure 7.5. This shows the type of chip technology employed, the number of components necessary, and the final selling price for a computer manufacturer wishing to supply 1M byte (i.e. 1 million characters) of primary memory, using the latest generally available RAM technology. As can be seen, over the past six years the

Shared Costs	wafers
	photolithography (masks)
	multi-layer etching
	testing and break-out

| | yield = percentage of good chips |

| Individual Costs | mounting |
| | assembly on boards |

Figure 7.4 *Economics of Production*

packing density for memory chips has doubled every year, so that the number of chips required to make up a megabyte has halved every year (with proportionate improvements in reliability). At the same time the final user price has fallen 10-fold, to less than 1 cent per character stored.

As an example of the switch from one technology to the next, consider the situation in mid-1981 when most computer companies were considering moving from using 16K-bit RAM chips to form their primary memory to using 64K-bit chips. At that time 16K chips cost slightly less than $2 (in large quantities), whilst the new 64K chips cost about $10 — thus on a per bit basis the 16K chips were still cheaper. However, use of the 64K chip

	Bits/Chip	Chips/M byte	$/M byte
1975	1K	8192	95528
1977	4K	2048	37198
1979	16K	512	14924
1981	64K	128	6100
* 1983	256K	32	2532
* 1985	1000K	8	1058

Figure 7.5 *Cost per Megabyte of RAM Primary Memory*

Note: * Data for 1983 and 1985 is predicted on present known technological developments

reduced asembly time and assembly costs for a given sized memory, and improved product reliability (fewer chips and board assemblies). As the computer companies wished to get on the 'learning curve' for using the new chips, many of them switched to the new chips for any new memory products launched during 1981, even though on a cost per bit basis they were still more expensive.

Given these economic and production characteristics of the chip, how has the semiconductor industry developed? And what are its dominant characteristics?

Historical Development of the Semiconductor Industry

The original stimulus (and the massive R&D funds) necessary for the manufacture of the first microelectronic components came from the American military authorities, particularly for missile control and guidance systems, where physical size (or more important the lack of it!), low power requirements, and reliability were paramount. Initially systems made up from individual transistors were used, but the development of the capability to manufacture integrated circuits led to an explosive growth in their subsequent development and application. Roughly the same period (the 1960s) saw major growth and developments in computing, and in particular the widespread application of computers in commercial data processing. The rapidly growing defence and computer industries formed the major markets for the products of the semiconductor companies until the beginning of the 1970s. By then the industry was on its by now familiar pattern of explosive growth in output and reduction in cost.

The original companies in the semiconductor industry, and still amongst the major ones, were those engaged in instrument-ation and precision engineering, such as Texas Instruments, Hewlett Packard, Fairchild and Motorola. Subsequently they were joined as manufacturers of semiconductor devices by most of the computer companies, and by several totally new com-panies, usually founded by people leaving one of the original manufacturers and setting up to exploit their own engineering or design flair, such as Intel and Mostek. Throughout the 1970s the semiconductor industry has been characterised by its dynamism, competitiveness and entrepreneurial flair, especially in the creation and exploitation of totally new markets for their products.

The market leader in the semiconductor industry is undoubtedly Texas Instruments (TI). Following TI come five companies of roughly equal size in terms of chip production, namely Motorola, Intel, National Semiconductor, Fairchild and Hitachi. TI is not only the largest company in the industry, it is also probably the one which has most successfully vertically integrated its operations into the higher value-added final customer products, such as calculators, watches, toys, computer terminals, mini and microcomputer systems — all major users of its integrated circuit output. The industry is also becoming increasingly dominated by the larger companies. In 1980 it has been estimated that more than 80% of production came from the top 10 manufacturers. The higher start-up costs, capital equipment and research budgets necessary to stay abreast of the field probably mean that INMOS (the UK semiconductor manufacturer founded by the NEB with government support) may well be the last major new entrant to the industry.

For new entrants much richer pickings appear to be available in the marketing of products and services utilising the technology, rather than in their manufacture. The semiconductor companies mentioned here are those who are in business to sell (*merchanting*, in the jargon of the industry) their semiconductor product in the market place. Many other companies, notably the computer manufacturers and those in electronic instrumentation and control also manufacture integrated circuits, but primarily (or often totally) for their own use. Industry estimates suggest that about 20% of total chip production is accounted for by in-house suppliers. Thus IBM is one of the largest (if not the largest) integrated circuit manufacturer, but its production is entirely for its own use. However, the computer manufacturers remain major customers for the products of the semiconductor companies (particularly for memory chips).

Vertical Integration and Value-Added

Many of the semiconductor companies have exhibited a strong desire to vertically integrate into the end-user markets for products and services using microelectronic components, in order to benefit from the greater value-added to be found in these markets, as compared with that obtained from simply manufacturing chips. Chips themselves are not sold directly to the final customer — they only reach him indirectly, and often in a form whereby the final customer doesn't realise (or even care) that he is purchasing microelectronic components. The customer is purchasing the final product for its performance characteristics — as a car, washing machine, robot, television or whatever.

The real value to the final customer lies not in the chips them-selves, but in the enhanced performance or cost-effectiveness of the product in which they are incorporated. The value-added in these final products is much higher than in the production of chips — this reinforces the desire of the semiconductor com-panies to enter some of these markets. The other major reason is to develop markets in which they can sell chips indirectly to final customers fast enough to keep up with their exponentially increasing output. If it had been left to the original manufacturers in the calculator and watch markets they wouldn't have developed the markets fast enough to absorb the vast quantities of chips that they in fact did absorb.

For the semiconductor companies and the companies using chips in their products, the world looks something like Figure 7.6, where the area of each part of the diagram is roughly propor-tional to the work done and value-added at each stage in build-ing up to the final system. Although this diagram was originally formulated in the context of computer system applications, it does have equal relevance to the incorporation of microelec-tronics in products. If we examine Figure 7.6, we note that the microprocessor with which we are concerned comprises the small top right-hand triangle — the microprocessor is an extremely small proportion of the final cost of most systems. To this are added some memory and logic circuits (typically to handle the input and output of information to and from the processor), and it is then packaged, perhaps only on a printed circuit board or in a rack, together with some power supplies to give the system builder an operational computer.

However, we still require several other items in order to have a system or product useful to the end-user. We need peripherals to communicate with the outside world — sensors, actuators, keyboards, displays and so on — and in the case of information

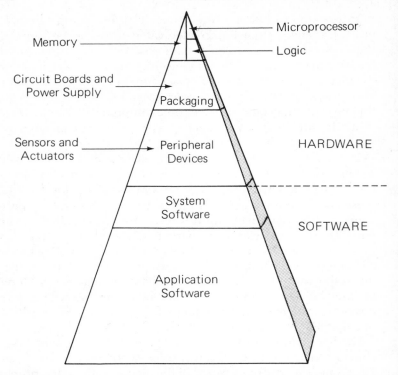

Figure 7.6 *Relative Costs in Producing Microelectronics Based Systems*

processing applications, some form of secondary memory device is required, such as a disk drive. This would now constitute the complete hardware of a computer system in the conventional sense, as well as in the context of other applications. In an electronic game, for example, the input unit may be a joystick and/or keyboard, and the output units may be some form of display together with a sound generator. However, in order to be able to respond meaningfully to the input from the user, the system requires some software, i.e. instructions on how to operate on the information input.

As we saw in Chapter 4, software is basically of two types: system software comprises the instructions enabling the processor and its associated components actually to function as a computer system (for example, telling the processor what form the input and output signals will take, or how to write and read data to and from secondary storage devices). System software of this type is often common to many applications of a particular processor, and is usually supplied by the manufacturer of the hardware. Application software is the specific instructions telling the processor how to deal with the data input in a specific application. In a general-purpose computer system these instructions will vary from task to task and will be loaded into the processor as required. In dedicated applications, such as calculators, watches, machine controls etc, they will probably be manufactured into a part of the memory of the unit. Thus the same 4-bit microprocessor may appear in a calculator, watch and game — the instructions in its associated memory will tell it in which product it is actually housed. Finally, there is the operation and use of the final system, which in a computer system context may well be the largest cost element in the entire operation. All the way down the triangle shown in Figure 7.6 the value-added in general increases, obviously not the same for every application, but usually following this broad overall pattern.

In evaluating where the value-added lies in a particular product utilising microelectronic components, it is often helpful to look at the ratio of the cost of the hardware items in Figure 7.6 (perhaps also including the system software) to the final market price of the end product. For example, if one examines this ratio for pocket calculators it is of the order of 70–80% — margins have now been pared down to the absolute minimum. However, for many electronic games (often using identical chips to those in calculators) the ratio is only half of this figure — the margins of the toy companies are much larger at present. To a large extent this reflects the stage in the product lifecycle that these two applications have reached. Electronic calculators are by now in a mature market with well defined market characteristics in terms of price/performance, and only a few large manufacturers. In contrast, the electronic games market is much younger and in a much more diffuse and varied state, with many producers (of games, not the hardware).

Growth in Output and Decline in Price

The dominant economic phenomena in the semiconductor industry are its exponential growth rate in output, and its continuing decline in unit cost. Looking at economies of scale in general over all industries, a rough rule of thumb would be that for each doubling of cumulative output we would expect a 20–30% reduction in unit costs. Examining the semiconductor industry we find that its output (in terms of electronic circuits) is doubling slightly faster than every other year, and that the costs of integrated circuits have declined by at least 25% per year for the past two decades. The cost of a given electronic function has been declining even more rapidly than the cost of integrated circuits, since the complexity of the circuits has been increasing as their price has decreased. Thus the comparatively straightforward regular logic memory chips have experienced rates of unit cost reduction approaching 38% per year over the past decade.

Such a pace in industry growth and unit cost decline is unique. This has been deliberate policy by the semiconductor companies who have followed what is known as 'learning curve pricing' — the quickest way to get down the learning curve to (hopefully) profitable high volume production is by aggressive pricing to develop new markets as rapidly as possible. The price to high volume customers for the next generation of chips is always below the current price on a per circuit basis, and is often below current production cost. Although other industries have shown evidence of similar economies of scale, they have done it by doubling output over decades rather than years. The result is that each reduction in unit cost (usually accompanied by a parallel increase in performance) opens up whole new application areas, generating ever higher output, technological advance, and lower costs. Rather than serving a single market growing at the same rate as GNP or total population, the industry is generating an ever widening set of markets.

However, there are some signs that the long fall in semiconductor prices may be slowing down. To a large extent the price of a general-purpose component (such as memory chips) has been fixed on a market-clearing basis. In the past the continued drop in prices has been necessary to open up new markets to take up the increased output, which in turn has been driven by the need to continue to exploit the considerable economies of scale in production. In 1979 things changed, in that for many components demand had outrun available supply. During the previous recession in the American economy in 1974, many of the semiconductor companies were caught out with excess capacity and large stocks, so that in anticipation of a recession towards the end of 1979 production plans for that year were fixed on the basis of a slow-down in the rate of growth in demand, but in fact there was a slight increase over the 1978 rate of growth.

Supply problems were compounded by production problems at TI, and by the unexpected large-scale demand from IBM for memory chips (particularly 16K RAMs), as it failed to achieve the required yields on its new in-house 64K-bit RAM chip. IBM is thought to have bought about 20 million 16K RAMs and EPROMs externally between mid-1978 and the end of 1979, and about the same number of 16K RAMs alone in 1980. This significantly contributed to the serious shortage of memory chips (especially 16K ones), which enabled the Japanese suppliers to gain about a 30% share of the American market for memory chips. As a consequence, the semiconductor companies found little need to continue their price reductions, and indeed towards the end of 1979 and during much of 1980 several actually announced small price increases.

There are two other factors which may mean that in the longer term the rate of decline in semiconductor prices may be at a slower rate during the 1980s than that exhibited during the 1970s. The first of these is that the price of many semiconductor devices has now fallen to such absolutely low levels that they are no longer a major or significant cost item in many of the final products in which they are incorporated. Second, the devices presently being developed by the industry are extremely and increasingly complex, both to manufacture and to use. In manufacturing, limits to the present optical techniques are being approached and the successor techniques will be very expensive to develop into mass-production systems. In use, customer companies are coming to expect much more (expensive) software and service support from the manufacturers.

As a consequence, further reductions in price do not necessarily open up new markets for the semiconductor industry. The development and marketing of new systems using the cheap intelligence that microprocessors provide is much more likely to do so. Thus, in the future the semiconductor companies may well put relatively much more effort into software and systems design. However, these *caveats* should not detract from the vast decline in prices that has already occurred; and remember that we are only referring to a slow-down in the rate of decline in prices — in real terms semiconductor devices will almost certainly continue to become cheaper and cheaper throughout the 1980s.

R&D and Innovation in the Semiconductor Industry

In an industry whose product declines in price by 25% per year the motivation for undertaking research and development is obviously very high. An R&D programme giving a company a year's advantage in launching a new product can give it a 25% cost advantage over its competitors. Similarly, a year's delay can put it at a severe disadvantage. Each new product is first made available to potential customers in prototype form, with mass production to follow as much as 9 to 15 months later. As early as this prototyping stage, the manufacturer is also expected to quote fairly firm prices at which the product will be eventually available (in lots of up to 100,000 or more!), prices which in the very nature of the industry are typically at least 25% below current prices, and are usually below current production cost.

The manufacturers of a new processor chip design also enter into what are known as *second-sourcing* agreements with one or more other semiconductor companies. Basically, this means that the originating company provides a set of chip production masks for another semiconductor company, who can then make the chip under licence. This reassures the customer that the design can be mass produced by another company, thus providing some guarantee of supply in the event of difficulties at the original manufacturers.

For all companies in the semiconductor industry, continuous product development is a fact of life, R&D expenditure typically takes at least 10% of revenue, and rapid product obsolescence is the norm. In the light of these comments about pricing policy, it is interesting to note that the recent problems of some semiconductor companies have been in the area of cash management rather than circuit design. The rewards for successful R&D activity are enormous, as are the penalties for failure. Up to now the industry has been typified by young companies — either very successful, or very dead! With the levels of interest rates seen in recent years, one of the major concerns of the semiconductor companies has been how to finance their growth. Each new chip design is an order of magnitude more expensive to design and develop than its predecessor, and requires an order of magnitude higher capital spending to get it into mass-production. Thus capital expenditures for the semiconductor companies were about 10% of sales in 1975, but had risen to about 16% by 1980, and are expected to be about 20% by 1985. For example, a new chip production facility will cost about $40 million within an existing plant, whilst National Semiconductor has spent about $100 million on its new Greenock facility alone (its total turnover in 1980 was about $1.2 billion).

Ironically, there have been recent changes in the organisational structure of the (American) semiconductor industry which some commentators believe may have a detrimental effect on its long-term rate of innovation and performance. These concerns arise as a result of the recent spate of takeovers of major semiconductor companies by much larger (usually conglomerate) organisations. In the two largest in 1979, United Technologies paid over £160 million for Mostek, and Scholumberger (the French conglomerate) paid almost £200 million for Fairchild.

The reasons behind this spate of takeovers are perhaps twofold — first, as already mentioned, the capital requirements of the next generation of products are enormous, and have strained the finances of many semiconductor companies, though some companies such as TI and Motorola in America, and the major Japanese manufacturers, have been able to draw on the considerable financial resources of their parent organisations. Second, several of the major electronics companies (and electronics-based conglomerates) have become concerned about their future access to semiconductor technology, and particularly after the problems of supply in 1979, to semiconductor manufacturing capacity. For example, several major chip customers were placed in severe financial difficulties by the shortage of 16K memory chips in 1979; the delivery times for many products doubled, and one minicomputer manufacturer (Data General) turned in its first-ever loss. Some takeovers have been by non-electronics end-users buying technical knowhow as well as capacity. For example, General Motors already has the capacity to make about 15% of its semiconductor requirements for itself. This is backward integration by the chip customers, contrasting with the forward integration by the semiconductor companies mentioned earlier.

The reason why these moves are viewed with concern by some commentators is because of the potential loss of engineering and entrepreneurial flair when these relatively small, often very personal companies, are absorbed by much larger and perhaps more impersonal organisations. The present semiconductor industry in America is almost entirely the creation of a dozen individuals, whose engineering flair in terms of chip design and production techniques, and aggressive marketing have dominated the industry. Yet, because of the strong competitive thrust and the need to innovate, in conventional financial terms the industry has never been very successful. Profits have been pared by competitive pricing and the need for R&D, so that the average rate return for the industry as a whole is only about 7% per annum. The danger perceived by some commentators is that the incorporation of the small semiconductor companies into big business will mean that corporate finance questions become more dominant, and the personally creative drive of these companies is lost. As a consequence, design leadership may well be lost (probably to the Japanese who seem best at corporate innovation), which in turn could lead to loss of industry leadership. In this sort of scenario the prospects for INMOS look to be quite good. Although the rate of innovation in the semiconductor industry may slow down for reasons such as these, enough has already been accomplished to ensure that a revolution will be wrought in most of the products produced by the rest of manufacturing industry, and in the provision of services to the community.

And so it is to the applications of this technology that we turn our attention in the final Chapter of Part III.

8

Applications of Microelectronics

In this chapter we will be reviewing the application of micro-electronics, and of the microprocessor in particular, in products. We will exclude from the discussion those products which come from the traditional computer and telecommunications industries, as we will be looking at them in greater detail in Parts IV and V. From the point of view of the semiconductor companies, the whole point of the markets for the application of chips is that they must be big — preferably very big! — and the primary means for stimulating and developing application markets has been an aggressive pricing policy. If necessary the semiconductor companies themselves have been willing to develop new markets if the existing firms in the market have been slow or uninterested.

Personal Products

Products for individual people obviously represent a very large market. The difficulty lies in discovering the kind of product that everybody will want. The two best known examples are digital watches and the electronic pocket calculator.

Calculators and Watches

In both instances the original firms in these industries appeared blissfully unaware of the fate that was about to overcome them.

It was the semiconductor companies, hungry for mass markets, who were the leading revolutionaries following the familiar pattern of aggressively pricing a technically superior product to achieve maximum volumes as rapidly as possible. For example, TI has by now sold over 20 million of its basic 4-bit processor, originally developed for use in calculators, but now used in many toys and games.

Electronic calculators were first introduced as desk-top machines in the mid-1960s at prices in excess of £1,000 each, but even at those price levels, their advantages over the then current mechanical or electro-mechanical calculators was so great that within a very few years they had totally swept the market. The availability of a technically superior, cost-effective product led to a rapid increase in demand. This in turn led to more technical developments in chip design (meaning more calculator functions for the customer and a marketing edge for the manufacturer), a better product, ever more demand, higher output, lower prices, more technical developments (in the manufacturing process as well as calculator capabilities) and so on, stimulated and fuelled by the aggressive marketing stance of the companies initially involved (largely Japanese). The customer as well as the manufacturer (of the calculator not the chip) benefited from vast increases in productivity. The price/performance improvement in the product has been dramatic — today's £10 calculator has a better specification than the £500 one of 1967, yet in the intervening years other consumer prices have more than doubled.

As a consequence, on average every household in the industrialised world possesses at least one electronic calculator. Although there were UK manufacturers of mechanical calculators, they completely missed out in the electronic calculator revolution, and this particular battle was fought out between the Americans and the Japanese. The original American calculator market in the mid-1960s was naturally dominated by domestic producers, yet by 1970 Japanese manufacturers of electronic calculators had captured about 40% of this market (by value). The Japanese share of the market peaked at 45% in 1971; but by 1974 it had fallen back to only just over 20%, whilst between 1970 and 1974 the American calculator market as a whole had more than doubled. What happened was that the major American semiconductor companies (particularly TI, Rockwell and National Semiconductor) entered the calculator market, vertically integrating to obtain the higher value-added from the final consumer product. Using their manufacturing base rapidly to lower prices and dramatically expand the market, they reaped the benefits from the economies of scale of mass-production.

The attraction of this type of product for the manufacturer is that it is a market obsessed by functionality. Each new version of a product has more functions, often for reduced cost. This is witnessed by the fact that you are probably on your second or third generation calculator. However, it is a difficult market to forecast. Fifteen years ago how many people would have seriously believed that nearly every person in the Western world would now have an electronic calculator? Even if they had believed that it was technically possible, they would surely have asked what these myriad calculators would be used for. Well strange to tell, nobody knows what they are (or probably aren't) used for, but people are still buying them.

Toys and Games

The other consumer area where microprocessors have already found widespread application is in *toys and games* — a prodigious consumer of chips. TI, the market leader in 4-bit microprocessors, also vertically integrated into the toys and games market (estimates suggest that in 1978, about 70% of TI's output of 4-bit micros went into toys and games), as well as terminals and minicomputers. For example, TI has sold over 7 million 4-bit processors for use in the 'Simon' game, with a one-time software development cost. As well as TI, the old-established toy and game companies have also leapt on to the bandwagon to the extent that roughly a third of 1978's entire microprocessor chip production went into this area. The big toy firms have seen their world revenues from these products rise to between 28% and 45% of their total revenue (from almost zero four years ago), apparently with relatively little effect on their existing product lines, and all of them seem to believe that the best is yet to come. In 1979, the two major toy companies, Mattell and Milton Bradley, together were the world's biggest purchasers of 4-bit microprocessors. Since this time the development of new markets has meant that the toy companies no longer have such a dominant position.

Some of what in this context are termed toys are perhaps

better viewed as being the first products to enter what will probably be one of the largest and most important markets of all: education. 'Toys' which make learning fun (for adults as well as children), such as TI's 'Speak and Spell' and 'Dataman' (a maths drill and practice device), are the tip of a huge iceberg of teaching systems, which can be personalised to allow the student to proceed at his own pace through the material (presented in an exciting fashion), whilst accurately monitoring his performance. The American semiconductor companies have never been slow to exploit a promising market opportunity, and the sums of money that some of them are at present investing in the development of microprocessor-based educational systems clearly indicate that they expect this to be a major and continuing market in the years to come. Educational applications are discussed as a separate topic later in the chapter.

What Next?

So what are the likely future products for this market? They will almost certainly be for things that we don't currently need or use; but in twenty years' time these products will be indispensable. We can speculate on a couple of likely candidates.

First the electronic dictionary and translator which are already beginning to appear. Simply type in a word in English and out comes the French/Italian/Spanish/German translation. At present this is an extremely literal word-for-word translation, but one can certainly see the demise of the foreign phrase book. These devices will undoubtedly acquire electronic voices and thus help with pronunciation. It is possible that there could be a revival in the quality of spoken English so that pocket electronic dictionaries would become fashionable. Both of these products appeal to the semiconductor manufacturers because they will require large numbers of memory chips to store the appropriate vocabularies.

A second possibility is the electronic diary, which could store phone numbers, appointments, notes, etc. Already some of the calculator/clocks presently available have a primitive message storage capability, and at some time in the future one must foresee the personal phone that maybe could be linked into this diary. But enough speculation, let us turn our attention to some of the other enormous markets that are already being exploited.

Domestic Products

After people, the next thing there are 'a lot of' is homes. The home contains a large number of specialised appliances, many of them covered by the term 'white goods', such as devices to cook, wash, clean, mix, dry, polish, etc. It is reasonable to assume that any device containing an electric motor is ripe for the inclusion of microprocessor control.

Domestic Appliances

One of the first companies to try microprocessor controls was Servis with their top range washing machine. The objectives, apart from the marketing reasons, were to produce a control device that replaced the existing electro-mechanical one, which was inflexible, somewhat unreliable and relatively expensive. This was achieved, though not without some considerable pain, and the resulting product had some additional advantages. First the mechanical parts of the machine (motor, drum, etc.) could be treated more kindly, for example by having control over starting-up procedures. The general state of the internal parts could be continuously monitored and warning lights indicate the need for maintenance. Maintenance itself could be easier with self-testing features built into the product. Greater levels of safety were possible, and finally the washing program was far more flexible — all this while designing a product that was fundamentally cheaper to build.

Subsequent to this pioneering work, many suppliers now have at least the option of microprocessor control on their microwave or ordinary ovens, washing-up machines, mixer/blenders, sewing machines, and soon no doubt on fridges, vacuum cleaners, electric drills etc. In a few years we expect that the term 'microcomputer controlled' will no longer have the same marketing appeal and will be quietly dropped. But before then we will probably have to live through the horror of all our appliances talking to us!

Environmental Control

Another major application in the home is controlling the heating, air-conditioning, lighting and security. The market for central heating controls is already being developed. They function in rather a clever way — you set the temperature required at a particular time and the controller will estimate how long it will take to warm up the house from the current ambient temperature. Obviously it cannot know exactly how long it will take, so it measures how far out it is and then uses this feedback information to recalibrate itself for the next day. Thus once the controller has experienced its environment for a few days it can then perform in an optimum fashion. Apart from the increased control over one's environment, this degree of control should reduce energy requirements. There can be a considerable saving, especially in office or public premises where most applications to date have been carried out.

Already in the US one can buy special digitally-coded switches for lights or almost any other electrical appliance. These digital switches can be controlled centrally by a personal computer for example. However, there can be rather amusing side effects, since the digital switching code is actually sent down the mains circuits, and consequently will travel to one's neighbours on the same circuit. Neighbourhoods where a number of people have installed these devices can observe such curious phenomena as all the garage doors in the road opening simultaneously, or lights in the house going on and off for no good reason.

It is likely that in the future a main service control computer will be built into each home as a standard component, which hopefully might give the consumer some control, or at least information, about usage of gas, electricity, water, etc. This is one view of the home computer, albeit somewhat different from the conventional one. We will be discussing other types of home and personal computers under the heading of leisure products.

Producers of this type of product have not been displaced in the same way that the watch and calculator manufacturers have been by the semiconductor companies. The microelectronic components typically only provide improved control functions, and represent but a small proportion of the total cost of the product.

Leisure Products

As we are being made painfully aware, leisure is becoming an occupational hazard. The demand for aids to help cope with the problem is on the increase, and this is an area in which the application of microelectronics can really begin to demonstrate some of its virtuosity.

Hi-Fi and TV

Anyone who has bought any hi-fi equipment in recent months will realise that microprocessor control is becoming a standard feature. The big question for most *aficionados* is, 'when will the whole system go digital?' The major experiment into domestic digital playback will come in 1982 with the launch of the Sony/ Philips audio disk. This is a small version of the video disk using the same laser technology, specifically designed for the storage of audio information in digital form. Though plans are well advanced, the major current limitation is in producing sufficient high quality disks at the right price. The record companies are enthusiastic as it is a playback only system. The facility for digitally recording audio information on tape is also possible but at present is being stopped by agreement between the manufacturers, though the need for it would not necessarily be great until digital broadcasting becomes widely available.

All of the limelight at the moment is falling on video systems. The colour television has reached a plateau in its development, and in the future two divergent streams can be seen at work. First, there is a desire for smaller sets that are not such a dominant feature of any living space, and secondly, TV projection systems are becoming very popular in the US. These can blow up the picture to about four feet square and project it onto a wall or screen. There is also a debate about what future form the TV will take. It may be simply a screen with very little electronics that can be plugged into a communications network in any room, thus accessing a main electronic control box. However, the flat solid-state screen would still appear to be many years away — currently it is possible to make a black and white flat screen which is about 10 inches square, but it would cost more than twice as much as a large colour television even in high volumes. Doubtless the electronic wall as portrayed in Ray Bradbury's *Fahrenheit 451* will arrive one day, but not this century.

Video

Currently, the hottest activity with video is in recording and playback systems. Video cassette recorders are an exploding market, despite the manufacturers doing their best to confuse potential customers. There are three different (incompatible) recording standards, so that potentially all the software (i.e. films etc.) has to be copied onto three different systems. For a long while it was considered that unless a common standard could be agreed, then VCRs would not take off in a big way. This has proved wrong and the television rental companies are finding a new lease of life for their high street shops. Unfortunately this welcome boom provides these companies with a somewhat vicarious pleasure, due to the uncertainty of developments in technology. They are being forced to tie up large amounts of working capital in both hardware and software when there is a distinct possibility of technology becoming obsolete well before the investment is paid back.

One technical development waiting in the wings is small portable video cameras which contain a mini-videocassette capable of recording thirty minutes of film. The film can be played back by putting the mini-cassette in a special dummy cassette that can be played on an existing domestic VCR. This has recently received a major boost from the agreement of all the manufacturers to standardise on a common tape recording format for this application. Does this spell the end of the cine camera, and/or will it lead to a great upsurge in home movie-making?

Still photography is not immune from this technology either. Sony has announced the Mavica, a camera that records images electronically and can be played back on a television set. In traditional photography, totally automative rangefinding and exposure setting are now standard features. This is done using a bat-like system where the microprocessors listen to the echo from an ultrasonic base.

However, this is surely a side show compared with the main arena where topping the bill is an all-out contest between VCRs and video disk players. The video disk is yet another technology with at least three totally different recording systems and standards. The technique pioneered by Philips uses a small laser to read the information stored optically frame by frame in digital form on the disk. This technique allows very accurate reproduction (video and sound) and direct access to any individual frame. The cost of this equipment is relatively expensive (currently over £500) but there is a greater problem at the moment with difficulties in manufacturing the video disks to the required standard.

In the US RCA have developed a different system called Selectavision. This is based on electrostatic principles with a stylus that actually touches the disk. Its advantage lies in the relative cheapness (about £300) and that the disk manufacturing process is under control. However, compared to the Philips Laservision, the disks are much more fragile and the picture quality inferior. It also cannot be used in a frame-by-frame mode. How will these machines affect VCRs? In America, RCA Selectavision equipment is selling very rapidly, and people are now speculating that the much forecasted demise of the cinema will now occur as new films can be distributed directly to the home. Film makers are excited because this is a playback-only mechanism and will continue as such for quite a few years. Disks are potentially much cheaper than tapes, as the software can be transferred in one single direct printing process.

But this is painting too rosy a picture of video disks. Outside the US there is considerable scepticism that they will have a significant impact on VCRs. While most people use a VCR to watch prerecorded films, it is its ability to act as a 'time shift' machine, allowing viewers to watch programmes at their convenience, which gives it a significant edge. As a part of the war between the VCR and video disk, Sony took the unprecedented step of a full-page negative advert in *The New York Times* with an obituary notice for the domestic video disk.

The future for video disk probably lies in its ability to access information frame-by-frame. Connecting a video disk to a microcomputer creates a very powerful and versatile training aid. For example, prototype educational systems are already available in which a video disk player is controlled through a program running on an Apple personal computer. In an industrial training context, General Motors are using a system like this to provide maintenance training and reference for their distributors. The customers can also look at the new models with the same system. If such a device were available in a large number of homes, then it could be used to distribute any of the catalogue information we now receive. For example mail order catalogues, holiday brochures, Yellow Pages and telephone directories and a whole host of educational material could be made available this way. The video disk is alive and well, though its big future will almost certainly not be as a direct rival to VCRs. It may well replace the traditional Computer Output to Microfilm (COM) systems, and is highly likely to be a basic technology for mass data storage on computer systems, being the likely successor to magnetic forms of data storage.

Industrial Products

Instrumentation Sensing and Control

One area where there has already been enormous change as a result of the impact of microprocessors on product design and performance is in those applications which come under the general heading of instrumentation sensing and control — in a sense the industrial equivalent of the developments in calculators and watches. Thus the instrumentation industry itself has had to redevelop almost its entire product range over the past five years, primarily making its products 'smart', introducing the era of truly automated measurement. One of the major manufacturers admits to having redesigned 60% of its product line in the past three years. This occurred at a time when the demand for the industry's products was rising rapidly, in parallel with the explosion in the wider use of microelectronics in general and microprocessors in particular.

From the customer's point of view, the new smart instruments automatically carry out the functions of the technician or laboratory assistant in running a test and analysing the resulting data. In 1978 it has been estimated that 40% of instruments sold incorporated at least one microprocessor, and this is an industry where sales are growing at a rate of almost 25% per year. The new instruments are easier to use, do not require degree-trained operators, incorporate facilities such as self-checking and diagnosis, and carry out a much wider range of tests in a much shorter time. For example, Hewlett Packard, the market leader in instrument sales, introduced a new machine which will carry out over 40 tests on telephone equipment, fully automatically, and in a few hours rather than the days it would have taken previously. The machine is controlled by a 16-bit microprocessor, and sells for about £10,000 — again an indication of where the value-added lies in microprocessors.

Robots

The word robot was coined by the Czech Karl Capek in his play 'RUR' (Rossum's Universal Robots) in 1920. In Czech the word 'robota' meant forced labour. Since then the word has come to have some quite specific connotations. Science fiction writers have explored the possibilities and none more than Isaac Asimov with his 'laws of robotics'. A robot is some kind of general-purpose programmable machine that can replicate human actions. It should be distinguished from a programmable machine tool which has quite specific (unhuman-like) functions. A robot is best defined by looking at the kind of robots that have been developed so far.

The majority of the world's 22,000 robots (estimated by the British Robot Association) are what would be termed 'first generation' machines — relatively crude, *un-intelligent mechanical arms,* performing some fixed repetitive task, often in conditions that would be dangerous or unsuitable for a human being. The most common applications are in paint spraying and arc-welding (usually of car bodies, as in the Fiat advertisement for the Strada). Other applications are in handling castings from furnaces, to feed metal presses, or to take products from moulding machinery. In all of these applications the robot blindly follows a set programmed routine, but has no awareness of the environment in which it is operating, or of the material that it is manipulating. Such robots typically cost between £25,000 and £60,000, but even at these prices and with their limited capabilities they are finding increasing applications in industry. The reasons are not hard to find. For example, a typical robot for welding car bodies will cost about £50,000; if we assume two-shift operation and a five-year life, then this works out at about £2.50 per hour — the present rate for car assembly line workers is about £6.50 per hour.

These first generation robots have been described as 'one-armed bandits — blind, daft, dumb devices screwed to the floor'. The simplest of these has been the pick-and-place variety capable of moving an object from point to point. Another class of robot specialises in a particular task, e.g. paint spraying. However, the trend is towards general-purpose robots. The world's leading manufacturer of robots, Unimation, has developed the PUMA (Programmable Universal Manipulator Arm) which is an arm of human scale capable of replacing people in existing production processes.

The main problem for people working with robots is that the robots work to much finer tolerances and are currently unable to cope with any degree of disorder. Nevertheless several manufacturers have been successful with this combination, notably General Motors.

There are three different ways of programming a robot: first, by an operator directing the robot through some remote controls; second, by the robot being given an example of what to do by physical manipulation (e.g. paint spraying): third, the robot may be under true programme control with instructions stored in some library. It is only in this final way that a robot can be rapidly made to switch from one task to another.

The next generation of robots will show three significant areas of improvement — in *dexterity, sensory powers* and *flexibility* — all of which derive from the availability and application of microelectronic components. Improved control and manipulative circuitry (smaller, so it can be incorporated in the robot, and much cheaper) enables robots to handle and manipulate relatively small and delicate objects, and to assemble precision equipment (impossible with first-generation robots). For example, robots can now assemble electrical motors and alternators.

The biggest breakthrough may be in improvements in sensing capability — in particular, giving robots the ability to 'see'. This would be done by linking the robot to a cheap monochrome television camera (itself a product of microelectronics), and a pattern-recognition computer. Pattern recognition is an excellent example of a computer application requiring considerable processing and memory resources which becomes economically viable in a much wider area of application as a direct consequence of the reductions in the cost of electronic components. Such a robot would be able to identify particular components for assembly from a selection available to it. Finally, the programmability and thus the flexibility of robots will be greatly enhanced by the general application of more computing power — a robot will be easily reprogrammable to cope with a variety of tasks.

Until recently the UK had been very backward in both the development and application of robots. In 1981 British industry showed signs of waking up to this technology. The number of robots in the UK went up to 700 compared to 10,000 in Japan, 5,000 in the US, 2,300 in Germany and 1,700 in Sweden. More encouragingly this represented a very high rate of growth in application. Unfortunately though, this means we are having to import all these robots, as we have virtually no manufacturing capability. While there is a lot of enthusiasm in this area, the scale of research funding (e.g. the Department of Industry's £10 million over three years) is miniscule compared to some of our competition.

CAD/CAM

The robot is only one element in the factory of the future. Manufacturing a product involves a whole range of activities from conceptualising, through design, analysis, simulation and drafting; to materials handling, forming, cutting, joining and processing; to quality and inventory control. Computers have long been associated with design, analysis and simulation but often as a series of unrelated activities. Using a CAD (*Computer Aided Design*) system the process can be co-ordinated using a series of common files or a database. This allows for a more rapid product design (and redesign) where the proposed item may be visualised and simulated in operation, and estimates of production cost and materials requirements can be immediately available. Once designed, the necessary drawings can be produced on an automatic drafting machine.

However the drawings may only be necessary if there are people involved in the production process. Traditionally the relevant materials would be moved between a series of machine tools each with its own *numerical control* (NC) system. Linking these machine tools together with a data transmission network was called direct numerical control (DNC) and adding a computer to control the machine tools directly (and dynamically) was known as CNC.

Putting together a mixture of machine tools and robots under the direction of a central computer produces the basis of a *Computer Aided Manufacturing* (CAM) system. The only missing ingredient is a facility to move materials from machine to machine. This might be a transfer line if the production volume of the product is high, or by using robot-controlled transport a *Flexible Manufacturing System* (FMS) can be created. Using the FMS approach it is economic to produce small batches of products and to mix the type of product being produced at one time. There are, as yet, very few proper FMSs (less than 100 in total). As perhaps is to be expected, Japan is leading the way in this approach to manufacturing and the only UK example is a small company in Somerset called Normailair Garret which makes parts for the Tornado military aircraft.

While all parts of the CAD/CAM systems are now becoming commonly available, linking them together in one integrated whole still has to be done for each individual situation. Given the interest and development in this area, one can expect that complete integrated systems will be available off the shelf within the next few years.

As an example of what can be done with current technology (indeed almost 10-year-old technology) consider Shell's fully automated Luboil blending plant at Shell Haven (opened in 1978). Customers' orders for the South-East of England are routed to the control computer and sorted into delivery sequence. This plant can blend more than 250 different lubricants from 30 base-oils and 80 additives. The oils and additives are stored in bulk tanks and barrels and connected to one of the eight blending stations on the main blending floor. Each lubricant is mixed in a blending vessel that is carried from station to station by means of a robocarrier. This is a bit like a fairground dodgem car that follows wires imbedded in the floor like a hidden electronic railway. The computer instructs the robocarriers to move the vessels from station to station, it instructs the station to add exactly the right amount of the necessary ingredient, it directs the vessel to a mixing station and finally tells the robocarrier to discharge the vessel through a hole in the floor that hopefully leads to a holding vessel before being transferred to a tanker. (A major requirement in system testing was sawdust!)

The computer is in charge of the whole process from checking the stock positions, availability of blending vessels and holding tanks; it selects the product formulae and works out the most efficient production program (for example, blending a light oil lubricant before a heavy oil means that the vessel does not have to be washed out). To Shell, the advantage of this plant is threefold; first the quality of the blended lubricant can be controlled to a much greater extent; second, orders can be processed much faster; and third, it is a much more efficient production process, thus maximising the use of expensive equipment.

In the future it is likely that total automated factories will become common. These production units will be capable of automatically taking orders, scheduling the work, producing the product from a store of components and then storing the finished product in some automated warehouse.

The Automobile Industry

Virtually all of the 9 million or more 1981 model-year cars that arrived in the dealer's showrooms in America in the autumn of 1980 incorporated a sophisticated electronic microprocessor-governed engine control system. This will provide about $600 million worth of business for the semiconductor companies in 1981, and at least $1 billion worth by 1985. The cost to the car companies has already been several times this latter figure in research, design and testing work. This situation has been forced on the American (and Japanese) manufacturers by the need to meet two stringent government requirements, one specifying the maximum amounts of noxious emissions allowed (of hydrocarbons, carbon dioxide and nitrous oxide), the other a minimum Corporate Average Fuel Economy (CAFE) which requires that the average fuel consumption of the total car output of each manufacturer meet some minimum standard, or else suffer substantial financial penalties. The American requirements are set out in Figure 8.1.

	Noxious Emissions Hydrocarbon/Carbon Dioxide/Nitrous Oxide Gramms/Mile	CAFE Miles/US Gallon
1978	1.5/15/2	18
1979	1.5/15/2	19
1980	0.41/7/2	20
1981	0.41/3.4/1	22
1982	. .	24
1983	. .	26
1984	. .	27

Figure 8.1 *Exhaust Emission and Fuel Economy Requirements for US Automobile Manufacturers*

As can be seen, the maximum allowable emissions tighten dramatically in both 1980 and 1981. Achieving any one in isolation would be straightforward, it is the combination of the three that is so difficult. All of the car manufacturers came to the conclusion that the only feasible way in which the standards in Figure 8.1 could be attained was by full computer control of the car engine. Ironically, in practice the rise in fuel prices in recent years and the resultant move to smaller and more fuel-efficient vehicles on the part of the car-buying public has meant that all of the American car manufacturers have been able to meet the CAFE requirements easily.

To illustrate what is involved, we will briefly describe the system that is being installed in Ford cars. It is typical, as all three major American manufacturers arrived at essentially the same solution — the anti-trust laws prevented the companies working on joint projects or pooling research. In the Ford system there are seven sensors continually monitoring the engine: measuring internal engine pressure and external barometric pressure; the internal engine temperature; the temperature of the cooling water; the throttle position; the position of a valve regulating the proportion of exhaust gases allowed to recirculate into the engine; the crankshaft position and rpm of the engine; and the proportion of oxygen in the exhaust. All of these sensors are continually sending information to a microprocessor, which in turn is continually controlling four actuators: an electronic fuel-injection system controlling the fuel/air mixture in the engine; an electronic ignition system, which sparks under control of the microprocessor, adjusting to the load on the engine; a system to control the air-flow to the catalytic converter, to control its temperature and the rate at which it soaks up pollutants; and control over the proportion of exhaust gases recirculated into the engine, effectively cooling it internally. A large part of the expenditure of the car companies has been on the design and testing of these sensors and actuators, and of the control circuits which interface them to the microprocessor, rather than on the microprocessor itself. The basic set of electronic components is common to all cars in a manufacturer's range — the system is told the characteristics of the particular engine, the catalytic converter, and the car that it is controlling by data stored in a programmable ROM.

To date it looks as though Motorola (through both its Semiconductor and Automotive Products divisions) will pick up the largest share of this business (about 30%), followed by TI and National Semiconductor. Apart from Motorola, which is something of a special case, the other major component suppliers to the motor industry have gained very little of this business, and one or two must be in some danger of losing substantial parts of their markets. Of the American car companies only GM has acquired a significant in-house semiconductor production capacity (and it at present aims to make only about 30% of its own requirements at most), whilst of the European companies Robert Bosch has a substantial holding (jointly with Borg-Warner) in one of the smaller American semiconductor manufacturers.

A problem from the point of view of the car manufacturers is that these systems are not very exciting from a customer marketing point of view, yet have involved the companies in vast R& D costs, and in ongoing costs of as much as $300 per car. Thus at present, all the car companies are trying to exploit their investment in semiconductor technology and knowhow by developing systems with much greater customer appeal. First on the list are new dashboard systems and diagnostics. The new dashboard systems that started to appear on many American cars in 1981 will be controlled by another on-board microprocessor. The new systems will not only replace existing instruments with new formats — for example, digital speedometers, and electronic bar-charts showing the amount of fuel remaining — but will also introduce new information, some of which is already appearing in cars — for example, systems for calculating average speed and miles per gallon. In future, all of these (and probably much more) will be a part of an integrated system. Diagnostic systems will be a mixture of both on-board and garage-based systems, and in most cases the on-board systems will be linked into the new dashboard systems, as driver-information systems.

As a part of the engine control systems most of the important engine functions can be readily monitored, and warning of other relevant items — for example, brake-pad wear, low hydraulic fluid, bulb failure, and heating/air-conditioning systems. Information and warnings on items such as these will be displayed as part of the dashboard system to the driver, perhaps initially as a set of codes, but eventually in English. It is quite possible that we shall see systems whereby the car's on-board diagnostic microprocessor is connected to a garage system when the vehicle needs servicing or repairing, in which the latter will interrogate the car's system and tell the mechanic the tools required and the sequence of operations to be followed. Allowing for developments in in-car entertainment and communications, suggestions that by 1985 every American car will incorporate more than ten microprocessors, together with a vast range of sensors, actuators and displays, appear to be quite reasonable.

Education

Some of the products that fall into the category of games have a strong educational flavour to them. 'Speak and Spell' provides spelling practice with the device speaking out the word to be spelt and then checking the spelling. There are a number of maths drill practice games, for example 'Dataman' and 'Little Professor'. The characteristic of these games is that they make learning fun, which would seem an aspect sadly missing in many of our schools, judging by the antipathy and indifference of many school children. Parents have always been prepared to spend considerable sums of money on their children's education, as was witnessed by the Encyclopedia boom. It seems likely that the semiconductor companies will develop a wide range of educational games that may well supplement the existing system, not without some opposition from the existing establishment, who view this development as a potential loss of control.

Teaching machines gained a bad name through the inflexible mechanical devices that were developed twenty years ago. Today, by employing microprocessors the options are very different, and computer-assisted learning systems are becoming of age. Probably the most prestigious of these is Control Data's Plato system. This system, costing many millions of dollars to develop, is based on specialised, sophisticated terminals linked to a powerful time-shared central computer. Each terminal is very powerful in itself; with colour graphics screens which are touch sensitive, recorded audio and visual material that may be played back on demand, it provides nearly all the facilities an electronic teacher needs. A vast number of lessons have been developed by teachers for this system; in these students go through the work at their own pace, and the system accurately monitors their progress. All the experience with Plato has shown that children enjoy using the system and almost certainly learn faster than when sharing the resources of a single teacher for a large variety of material. The major limitation up to now has been cost, since Plato has to be paid for by the hour and it is too expensive to be a widespread facility. Recently, Control Data have announced Micro-Plato, based on a small computer which could make this approach affordable by many schools. Certainly this type of approach is being copied by a number of other companies, and the possibilities appear boundless when you consider linking a personal computer to a video disk.

Another aspect of education and the microchip is that of teaching students about the technology itself, the hardware and the software. Many schools in the UK have acquired micro-computers through their Parent Teachers Association. After several years of discussion the UK government's plans to assist schools have come to fruition. The Department of Education and Science (DES) have obtained funds to start a significant teacher training programme, and regional workshops and training centres are being established. To buy the hardware the Department of Industry is providing 50% grants for all schools that have not been resourceful enough to obtain a computer already. However, choice is restricted to one of two British manufactured micros; either Research Machines 380Z or the Acorn Atom. However, considerable discontent has been caused by there only being two authorised suppliers; for example, Clive Sinclair was sufficiently incensed to offer schools the same terms on his ZX81 (i.e. he would give a 50% discount) and he has now supplied at least two thousand units.

But the major problem lies in the incompatibility between the different types of hardware. Many schools have bought Apples and Pets solely because of the vast library of software available. Thus schools are finding that their allocated computer arrives with maybe one member of staff who has been on a course, but no pre-packaged programs available. The situation may well become worse in the short term with the arrival of the BBC micro. This has been developed under contract by Acorn to be used in conjuction with an educational series of broadcast programmes. It uses a very sophisticated design and there are good reasons for wanting extensions to BASIC, but it means that yet another piece of hardware will be available with no library of software. With an actual production target of 50,000 units this

would seem a very juicy market for budding software entrepreneurs.

Small Scale Computers

There are so many microcomputers marketed with different
labels that it is useful to have some form of classification. One
can identify three distinct classes, though there is obviously
considerable overlap. We have labelled the three, personal, professional and small business computers for convenience, and
examples of particular systems in each category are shown in
Figure 8.2.

Personal Computers

These typically consist of a keyboard, up to 16K of memory,
maybe an integral display screen, a cassette recorder, and
possibly a simple printer. In the UK these machines cost from
about £70 up to £1,000, and are suitable for hobbyists, self-
teaching, and schools; but they are really too small for any
serious data processing applications. For some curious reason
the UK has been very successful at producing the machines at
the low end of this range. The Sinclair ZX81 has been incredibly
successful after a slightly shaky start with the ZX80. It has sold

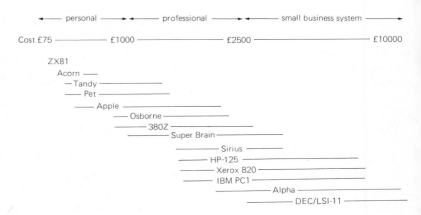

Figure 8.2 *Examples of Small Scale Computing Suppliers*

well over 500,000 units initially by mail order, but they are now additionally available through the W.H. Smith high street chain.

The Acorn Atom has been very successful, and though there have always been production delays, one hopes these will be overcome with the appearance of the BBC micro. Nascom have been technically successful, though poor business management led the company into liquidation from which it is currently being rescued by Lucas. Tangerine were the first company to bring out a low-priced machine with a Prestel adaptor; in addition, there are a host of others, some in production, others like the New Brain long awaited, others like the VIC-20 maybe ready to set new standards in the market.

In North America the TRS 80 was and still is the most successful personal computer. Its success is in many ways due to the distribution system that Tandy already had for its traditional electronic products through the Radio Shack Stores. Apple and Commodore (Pet) had to establish distribution outlets which took them a significant time. It is only now that Apple has hauled back Tandy's long lead through the general superiority of its product. Apple are however at the top end of this price category, and Commodore who have successfully occupied the middle ground, particularly in Europe, have now introduced the VIC-20.

Professional Computers

The line between this and the previous category is very blurred, and many of the machines belong in both categories, depending on their configuration. The hardware would normally consist of a keyboard and screen (possibly colour) with 48/64K of RAM memory, floppy disk(s), a reasonable quality printer and maybe a number of other specialist devices such as a light pen, graphics tablet, plotter, etc. In the UK such a system would currently cost between £1,000 and £2,500, and could be used for a wide variety of purposes; from education to small business accounting; from a house vending register to marine route planning; from word-processing to a patient record system; from CAD (Computer Aided Design) to financial planning.

They represent the major growth area in the sales of micro-computers. It seems that virtually any data processing application that can be usefully undertaken with a few thousand records of storage and a few hundred transactions per day can be accommodated on such systems. Within organisations, these systems are often being used as part of a distributed network. The network may be solely a series of linked micros, or the micros may be linked to a central mini or mainframe computer.

Until recently this market had been the preserve of the small specialised microcomputer companies. Up to now, Apple has been pre-eminent in this field, though their hardware tended to run out of steam at the top end. Their major competitors pursued a course, which may prove decisive in the long run, of using one of two microprocessors — the Intel 8080 (and its derivatives) and Zilog Z80. For this microprocessor a company called Digital Research developed an operating system called CP/M. Subsequently CP/M has been implemented on a number of other microprocessors, and a number of compilers (notably BASIC, FORTRAN, and PASCAL) have been developed by companies like Microsoft. The great advantage of such an approach is that software may be transferred between different microcomputers. Thus much of the world's micro software is moving towards a CP/M compatible standard.

This process is being hastened by the action of several of the major computing companies. Hewlett Packard, traditionally on the edge of the field in providing calculating power to professionals like engineers, brought out first the HP 80 and now the HP 125. This is a fully-fledged CP/M machine, running programs like an enhanced Visicalc. The second development was in mid-1981 when Xerox released their 820 microcomputer — one of the devices that make up their electronic office system range which may be connected together using Ethernet (see Chapter 10). However, Xerox's internal code name for this machine was 'the worm', which makes their intentions towards a rival product quite explicit! This again is a CP/M machine supporting a version of Visicalc and a word-processing package.

But possibly the most significant development was the announcement in the third quarter of 1981 by IBM of their own personal computer, PC1, again a CP/M-based machine. The IBM machine is also potentially a very different beast from its competitors in that it utilises a 16-bit architecture. If this potential is realised it will make this machine a serious contender in the next category. This has had the effect of legitimising the whole area of microcomputers within the traditional dp framework. It is interesting to see how IBM intend to market this machine; first through its own retail shops, second through Sears Roebuck and Computerland; but most important of all, through its own existing large-account computer salesforce. This indicates that IBM see a primary role of this micro as a terminal for their larger computers, and almost certainly expect the dp managers of their existing computer installations to be potent selling agents in maintaining single-supplier status in many organisations.

There must be a major question mark over whether the small specialist companies will survive the onslaught by these computing giants. It seems to us unlikely that those providing CP/M machines will have sufficient marketing edge to compete unless they can come up with significant technical innovation. However, Apple is sufficiently different and sophisticated in the way it handles graphics and colour that it is unlikely to be overwhelmed. It has also established such a significant position in the market place that the investment in user software is too high for many users to make a radical change.

Small Business Systems

The final category is for micros that are designed to handle the traditional data processing requirements of a small organisation or business unit. This would require potentially a number of terminals (keyboard and screens), more than 64K of memory, a small hard disk storing 10 or 20 Mbytes, and a good quality printer. A key concern may well be the provision of file storage backup. This may be achieved by having two disks, though the mini-disks do not have replaceable cartridges, so one of the new cartridge tapes may be required.

In many ways the hardware is much less important than the software and the level of support from the supplier. The software may be developed using one of the transportable operating systems or compilers like micro-Cobol from Micro Focus, Digital Research's CP/M, Microsoft BASIC or PASCAL. The key is a set of integrated business programs which often give better facilities than those found in comparatively larger organisations. Much of the hardware used is effectively small oneboard minicomputers such as Digital Equipment's LSI-11 or Data General's Micro Nova.

The distinctions between these categories is one of usage

rather than the hardware. In the next few years we can expect that the power of the hardware will increase by several orders of magnitude and the cost will drop substantially, but the way the hardware is used will most likely continue to fall into these categories.

The Home Computer

Is the concept of home computer realistic or is it a hopeless pipedream? This is the $64,000 question. Texas Instruments, in particular, believes passionately in its future and to this end has spent a fortune trying to develop it. But what would a home computer do? If we leave aside the control function aspect that we have already discussed, then there would seem to be four major application areas, home economics, leisure, education and making transactions. On their personal computer TI has gone to considerable trouble to develop attractive home economics programs. Household financial budgeting and cheque book reconciliation are, however, rather far removed from the average household, many of whom have no bank account and little discretionary spending power. Computerised cook books also seem to be scraping the bottom of the barrel somewhat.

Leisure would seem a more promising avenue and this is the direction the game playing machines have taken. The computer can become a very stimulating companion, but there are some worrying implications. When children play games, less than half the time is spent on actually playing the game, while the remaining time is spent negotiating rules and practising the arts of social behaviour. When playing against an electronic component, the skill of the game is everything. There is a burning desire to beat the machine, but no negotiating or social skills practice.

The home computer could become the heart of some central electronics complex providing hi-fi, TV, video recording and a number of other features. Recently Rediffusion Computers announced the first offering in this field. Called the Teleputer, it combines a personal computer with a video disk and VCR. This is not initially aimed at the home market but rather to be used as a terminal with private viewdata systems.

With increasing openness of society, the demand for enlightenment and education grows. The home computer provides a two-way interactive medium and this is a great advance over the TV as an educational mechanism. Whether each home computer could contain all the required information seems doubtful, though there is no reason why material could not be distributed on video disk through a lending library scheme. The other alternative is to access the information via a telecommunications link using a system like Prestel. Today the cost of such a system would be at least four times as much as a VCR, but this can change very rapidly.

But the main reason to be certain that the home computer will happen one day would seem to be its potential for making transactions and retrieving records of transactions. The main contenders here are the banks who would like to overcome some of

Personal Products	Calculators/Watches
	Dictionaries/Translators
Domestic Products	Control Systems for: Appliances
	Central Heating
	Security Systems
Leisure	Audio/Video Record/Playback Systems
	Toys and Games
Education	Drill and Practice Aids
	Computer-aided Learning (CAL)
	Computer Science
Small-Scale Computing	Personal
	Professional ⎫ Computers
	Small Business ⎭
Automobiles	Fuel Efficiency/Pollution Control
	Driver Information
	Navigation and Safety
Instrumentation	Multi-function Devices
	Built-in Processing Capability
Robots and Factory Automation	Mechanical Arms
	Sensing and Programmable Robots
	CAD/CAM Systems

Figure 8.3 *Application of Microelectronics — Some Examples*

their paperwork by allowing statements to be viewed electronically. A customer could also set up a standing order directly. The
ability to make transactions could be used by mail order companies, holiday operators, travel companies and others. Note
that the transactions are made with the service provider rather
than with their agents (with disturbing implications for the service industry). This facility can be effectively provided via Prestel
and the new gateway facility (one-stop electronic shopping as it
is known). We will be discussing the future for Prestel in Chapter 10.

All in all, it seems that the home computer will happen one
day. The cost of the hardware will need to come down to today's
VCR prices before it is likely to happen and then it may need the
confidence of the television rental companies to create the
essential critical mass.

Part IV

THE CONVERGENT TECHNOLOGIES

As we saw in Part III, microelectronics are having a major impact on many products and processes, and none more so than the information processing industry itself. This new industry is arising out of the convergence of three technologies — traditional computer data processing, especially in the development of distributed data processing; telecommunications; and office products and services. All three industries are converging to serve a common market — that of information processing. The computer companies are marketing ever cheaper products, yet with higher end-user performance, distributing processing power to the end-user. The telecommunications companies have discovered that future methods of transmission are more akin to how computers talk to each other than how telephone systems have previously operated. And the office products suppliers are being overtaken by a rash of 'intelligent' devices, much as the instrument makers have already been subjected to.

9

Distributed Data Processing

Information Processing Industry

The major operations involved in processing information are its generation or capture, storage, subsequent retrieval, processing, transmission and display. The computer companies have long been engaged in all of these operations on data, and find it logical to extend into the same activities for other information media — voice, video, text and facsimile, particularly as these media are increasingly digitised. However, the activities of the computer companies in their existing markets, especially the present promotion of the distributed processing concept, are increasingly involving them with telecommunications.

The market for telecommunications services in general is rising rapidly in all industrialised countries (and in many of the less-developed ones too), in all of which the services have traditionally been provided by a publicly-owned or regulated corporation. The telecommunications companies are well aware of the expansion in interests of the computer companies and are generally fighting back by moving into the wider markets of information handling equipment to hang on to their networks. At the same time the office equipment manufacturers, particularly those of typewriters and copiers, are being forced to increase the functionality of their products to compete with the alternative approaches being offered by the computer companies.

The point about this *convergence* is that although Intel and Motorola may produce chips that end up performing control functions in cars or washing machines, we don't expect them to set up in competition with Ford or Hoover. However, in information processing, companies such as IBM, AT&T and Xerox are all invading each other's territories, and even complete outsiders such as Exxon are entering the field. One reason is that in the case of the car or washing machine the incorporation of microelectronics doesn't change the fundamental purpose of the product — it simply improves its all-round performance and cost-effectiveness.

But information processing is in a sense a new market which until now has not been approached in a coherent and systematic manner. The incorporation of microelectronics is resulting in products whose performance has changed so much that they can be regarded as a completely new product. A computer-controlled, electronic telephone exchange or a laser-printing, computer-driven copying machine with telecommunications capabilities are very different products for their users, compared to their predecessors.

Approaches to DDP

Distributed data processing (ddp) is a term which is unfortunately very rich in conjuring up appropriate images in the reader's mind, but which lacks any formal agreed definition as to its meaning. Partly this is because in many instances it is a concept, rather than a reality; and partly because it has become a marketing necessity not only for the computer companies, but particularly for the vendors in the small business systems and office automation markets. Thus distributed data processing has been used to describe systems as diverse as:

>hundreds, in some cases even thousands of terminals connected to mainframes in a strict hierarchy, where the mainframes are master and the terminals slaves;

>a number of minicomputers linked together in a network as equals with terminals linked to each mini;

>a network of personal microcomputers, sharing resources such as disks and printers, and perhaps linked to a more powerful central computer facility;

>the installation of stand-alone personal computers on every manager's desk.

These different systems are illustrated schematically in Figure
9.1 (a)–(d). Thus diagram (a) illustrates the first of these alternatives, commonly known as a *star* system, which is the situation
to which the dp configurations in many large organisations have
evolved. Figure 9.1 (b) illustrates the second alternative, which is
typical of a minicomputer *network*, in which each machine may
be performing the same function (e.g. order-entry systems in
several sales regions), or may be specialised (e.g. one performing
order-entry, one manufacturing control, one general accounting,
etc.). The third alternative in diagram (c) is an example of what is
known as a *local area network*, which will be discussed in more
detail later. The last, diagram (d), is an example of what has been
termed *'fragmented'* computing.

The movement towards ddp through the 1970s has been fuelled by two critical pressures: the ever reducing cost of the
technology, and the increasing inability of the existing computer
systems to meet the real needs of the user. Both of these descrip-
tions are gross over-simplifications. They encompass related
developments which we now examine in greater detail; first, we
look at recent developments in computer systems.

Recent Technological Developments

The overwhelming development in computer systems over the
past two decades has been the reduction in cost of an entry-level
system, such that today it can be afforded by an individual; or
alternatively, the provision of greatly increased computational
capabilities for a given cost. Clearly one of the major reasons for
this has been the dramatic improvements in price/performance
brought about by the technological developments in microelectronics. As we have seen, progress in chip technology has
had an enormous impact on processor and primary costs. However, developments in chip technology have had their greatest
impact on the price/performance of the cpu, but what about the
price/performance of the major peripherals, particularly secondary memory and terminals? Here too there have been dramatic
improvements in price/performance, due largely to the rapid
growth in the industry, but also to the fact that the number of
peripherals associated with each cpu has risen over the years.

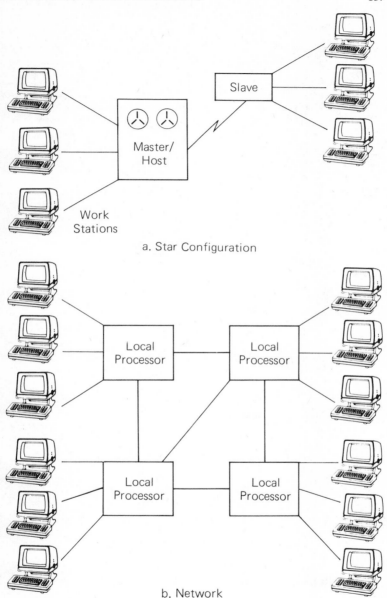

a. Star Configuration

b. Network

Figure 9.1 *Distributed Data Processing*

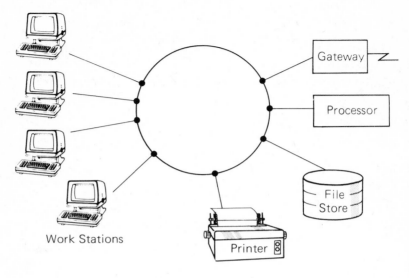

c. Local Area Network

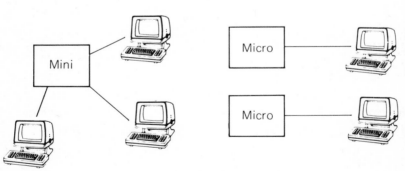

d. 'Fragmented' Approach

Figure 9.1 (continued) *Distributed Data Processing*

Disk Memory

The vast range of peripherals, particularly disk and tape drives, printers, and to a lesser extent, terminals, are relatively unaffected by the technological developments in semiconductor technology, since they are primarily mechanical devices. Yet even here there have been dramatic price reductions — Figure 9.2 shows the decline in cost for 1 Mbyte (one million characters)

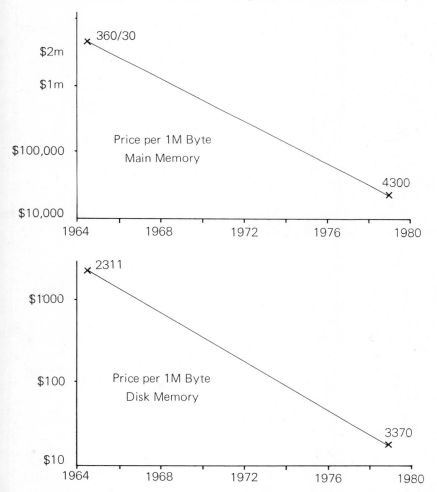

Figure 9.2 *The Decline in Memory Costs (Typically IBM Equipment is Cited for Illustrative Purposes)*

of main memory (which from the mid-1970s onwards has been semiconductor-based. It also shows the decline in cost for 1 Mbyte of disk memory.

Apart from the absolute differences in cost, both exhibit similar declines; yet disk drives (as opposed to their controllers) have benefited directly very little from the developments in semiconductor technology. The decline in cost of disk drives (and other peripherals) stems from the normal technological developments and learning curves in any product — in this case it has been accelerated and compressed by the rapid growth and fierce competition in the industry. As in the case of the semiconductor companies making standard memory chips, so for the companies making disks: both are to a significant extent manufacturing a product to meet a set of technical criteria; competition is on the basis of price/performance and not on brand loyalty (except perhaps for IBM).

Similarly, technical R&D is a fact of life — you must innovate just to stay abreast of the competition, and if you are lucky you may get ahead for a short time. Increasing amounts of storage can be placed in smaller and smaller boxes, so that even small entry-level business systems often come equipped with disk drives which have a capacity larger than any drive available 15 years ago, and in about 1/10th of the space.

Terminals

Progress in terminals has been spectacular in two ways: first, as usual, in price reductions (though, as the unit price is low, large percentage decreases don't perhaps have the impact that they might for other devices). Second is the ever-widening range of 'intelligent' devices available which are capable of being attached to a computer through some standard interface. The display terminal ranges from the basic KSR (keyboard send receive) terminal to a device, with local memory and processing capability, operating in glorious technicolor! (The dividing line between sophisticated terminals and microcomputer systems is blurred in many instances.) The printing terminal is adopting new technologies to provide higher speed, greater reliability and much improved quality. For example, the future may be with ink-jet printers, which spray drops of ink on to the paper to form the characters, giving high quality and a wide range of fonts under program control; or with cheap laser printers, in which a laser is used to 'paint' an image on to a Xerographic drum, as in a photocopier.

Terminal devices which a few years ago would have been regarded as too expensive or esoteric for commercial application, are now becoming commonplace: graphic devices, both screens and plotters, are widespread; voice output is in commercial use (for confirming sales orders) and voice input is not far behind (for restricted vocabulary systems); systems to read typescript are well developed, and one to read handwriting was announced recently — almost all incorporate a microprocessor. Not only have general terminal devices proliferated, but many special-purpose devices are also being developed using cheap micro-electronics, and many organisations are finding it increasingly economic to develop their own devices for their own specific applications. Retail point-of-sale terminals (i.e. sophisticated microprocessor-controlled cash registers, incorporating a bar-code reader, and communicating with a larger system for stock recording and accounting); fully automatic bank teller machines; airline ticket issuing machines; travellers cheque issuing machines; and many more — all are examples of specialist 'smart' terminals.

Economies of Scale

A consequence of these technological developments is that one of the 'laws' of computing has become increasingly redundant — if indeed it is still true at all. The 'law' in question is Grosch's law (named after its originator in the late 1950s, Herb Grosch), which basically states that the power (processing speed) of a computer is proportional to the square of its cost. That is, if computer B was twice as expensive as computer A, then computer B would be four times as powerful as computer A. These economies of scale were empirically observed during the 1950s, 1960s and early 1970s. Along with the associated costs of a controlled environment for the machine and the specialist expensive personnel required to tend it, this accounts for the almost universal implementation of highly centralised dp operations in organisations. The focus of management control was typically on cost per transaction processed and stored — this was usually minimised by processing everything through one big machine. It was primarily the impact of these economies of scale which forced the centralised dp department to try and be all (computing) things to all users.

However the decline in the cost of computing equipment gave rise to the possibility of breaking up this monolithic structure. Whether or not Grosch's law is still true today is almost irrelevant. As we have already seen, the cost of the hardware is a smaller and smaller proportion of the total cost of a computer system. Thus the economies of scale in processing cost obtainable by moving to a larger machine are less and less significant, and are increasingly outweighed by the disadvantages of centralisation. These factors may be summarised as follows:

> *lower relative cost* of the cpu as compared to the total cost of the system;

> *lower absolute cost* of processing on a per transaction basis because of improved hardware price/performance;

> *increasing flexibility* in application because of less stringent environmental requirements, smaller physical size, and increased reliability.

From a technological point of view it is now possible to tailor the computer system to the application, and not *vice versa*. And not just at the small end of the market either. One of the big success stories of the 1970s was that of the Amdahl company. Its founder, Gene Amdahl was the designer of the large machines in IBM's 360 and 370 ranges, and believed that there was a market for even larger machines. However, many commentators thought that the days of the large machine were numbered, so that Amdahl had to leave IBM and found his own company to design and manufacture a range of very large computers. His company was very sucessful in demonstrating that there is a range of applications which requires a very large machine in its own right. However, from the point of view of the economics of the available technology, there is now no longer necessarily any need for a centralised dp operation to attempt to meet all of the computing needs of an organisation — the computer can be economically tailored to the application. Note that this *doesn't* necessarily mean that all applications should be distributed — it may be the case that the economics of an application require centralised processing (airline seat reservation systems are a good example of an application requiring a central facility).

User Requirements

All that we have discussed so far are the technological developments which make some form of distributed processing economically viable if appropriate. In practice it is often the pressure of
users which has led to a distributed processing solution in many
organisations, particularly user resistance to any further applications being undertaken on a central facility. Through the 1970s
many organisations have been faced with increased disillusionment with, and resistance to, the centralised dp operation,
especially in those cases where their organisational structure in
other respects is not centralised. Several reasons have been
suggested for this. The most important ones are:

> stagnation or deterioration in the level of service provided,
> largely because of the inability of a centralised dp facility to
> be all things to a diverse group of users;

> having to accept some general, inadequate and inapprop-
> riate level of service, with little direct control, influence or
> priority selection rather than a service tailored to the user's
> specific requirements.

From the user's point of view, the general dissatisfaction with
the service provided can be particularly related to four specific
factors, namely:

> *Worsening reponse and throughput:* a shared system may be
> too busy for the really large jobs, too slow for interactive
> work, and in general too unresponsive to the differing
> priorities between jobs.

> *Poorer availability and reliability:* there is some evidence that
> in a complex shared system interruptions in service are
> more frequent than in a dedicated environment.

> *Increasing overhead:* the necessary complexity in the operat-
> ing system software of a shared system trying to be all
> things to all users, means that this software tends to con-
> sume an increasing proportion of machine resources, and it
> is the user who bears the direct and indirect costs of this
> (through the financial cost of the system, and of the
> increased resources necessary, or of reduced resources
> available to the user).

> *Increased rigidity and inflexibility* of procedures, priorities

and the working environment are frequently present in a highly centralised and structured operation.

There are also three major areas of concern to users in which distributed data processing can make a significant contribution.

The first is *data entry* — one of the major problems in computer application is getting data input accurate (in computer jargon this is called 'data capture'). In the early days all input data had to be first written on to special forms and then punched on to cards or paper tape via a keyboard. Two potential errors occurred; first in the initial recording, and second in the transcription. The transcription errors could be almost totally removed, at some cost, by the process of verification (i.e. punching the data twice and making sure it matched), but recording errors were less easily overcome. The major reason for this was that the originator of the data was separated from the data input process. Data preparation personnel had no knowledge of what was reasonable or sensible in the data and this was the cause of many of the stories about computer mistakes.

Distributed data processing allows the data capture process to be taken to the originator of the data. Very often this means that the recording and transcription can be combined into one process. Data can be captured as a part of an interactive application, for example, an on-line sales order processing system, where the input data is a transaction that has to be validated and processed before more data can be input. Many errors can be detected at this stage (for example, non-existent, incorrect customer number etc.) and corrected, because the user can make reasonable judgments about what the data should be.

The second area of concern to users is the opportunity to *reduce the load on the central computer*. Whilst central dp computers operated in a batch mode, work could be scheduled in an efficient manner and the load spread out over the available time. However, when an on-line mode was introduced, central dp was no longer able to schedule this work. For example, if a company made stock records available for enquiry, say between 10–12 a.m. and 2–4 p.m., then there would be a heavy volume of enquiries between these times, over which dp management has little or no control. The on-line requirements often increased the required computer power quite disproportionately to the amount of work being done, in that it was required only for limited periods of time. Another factor was the unsuitability of many early mainframes for on-line working. Distributed data processing was a cost-effective way of bringing power to the user.

Finally, there is the question of *ownership* — of the data and the processing resources, and in particular over scheduling and priority setting. As we have discussed above, central dp has become alienated from many users. This is an organisational problem more than anything else. Distributed data processing is a way of allowing users to regain ownership of their data processing capacity. This is a two-edged sword as it forces the users to be directly involved in the dp process, often for the first time.

So what is distributed data processing? Well it can cover a wide range of alternatives and it is important to distinguish between distributed (or dispersed) dp and decentralised dp.

Decentralised versus Distributed Data Processing

With *decentralised* dp, responsibility for both the planning, implementation and use lies with the user, and the only involvement with a central department might be as a consultancy service. The disadvantage of this approach is that there is likely to be incompatibility between the different processing units, or different standards for data processing or storage procedures, or communications capabilities. In some organisations this may not be a problem (e.g. groups with disparate divisions), but it can cause problems.

The classic case is Citibank who decided in the mid-1970s to decentralise their dp operation in what was called 'Project Paradise'. The idea was to 'let the flowers grow'; most central dp personnel were transferred to user departments and everybody went their own way. Certainly some remarkable developments took place, with divisions having special terminals and even special minicomputers built for them. But in the end the lack of compatibility between divisions became intolerable. Central dp was reformed to tie all the parts back together again. Certainly as a result of this project, the Citibank users were probably more knowledgeable and motivated about dp than in any comparable organisation. The missing ingredient, which was subsequently imposed, was a network which could link the different parts together, enforcing some common standards, and maintained by the central dp department.

On the other hand, *distributed* dp implies that the overall responsibility for the system still lies with some central department. The role of this department may vary from just defining standards and authorising types of system, to having total responsibility for system development and implementation (though physically the systems would be located with the users). This relationship will reflect both the type of organisation and the history of dp-user relations. Centrally managed organisations are unlikely to relinquish much control, but they do accept that it is sensible to put the processing capability where it is required. However, organisations with weak central management or where a central department has historically failed to deliver appropriate dp systems to user management, will find that ddp can happen very rapidly. This can become akin to a guerilla war, in which the user's new-found weapon is the personal/professional computer.

There are also quite different philosophies of approach proposed by the different computer suppliers. As described above, the mainframe manufacturers, led by IBM, have favoured a large central machine (called the 'host') with a number of intelligent satellites all under its control. On the other hand the minicomputer manufacturers favour networks of machines capable of communicating with each other as equals. This latter approach is becoming increasingly attractive with the emergence of both internal and external networks. The technical characteristics of networks are discussed in the next chapter.

A major problem with ddp can occur when one attempts to distribute data as well as processing power. For example, consider a distribution company with a central warehouse. If each distribution outlet has some processing capability with the current stock position available, then any update to the local data will need to be reflected in all the other outlets. A more practical approach is to load the stock availability figures into the local machine before the start of work and only update the local data during the day. Overnight the daily movements are transmitted to a central machine, which, once it has collected the data from all the outlets, can produce the new stock figures for the next day. This type of system is known as 'pseudo real-time', in that it appears to be presenting current information, but the information it presents will become increasingly out of date as the day proceeds. The acceptability of this type of system would depend on the rate of stock turnover and the perishability of the stock.

If the distribution company had regional warehouses which served specific outlets, then the data on stock availability in a particular warehouse could be maintained in a machine at that local warehouse. Communication between machines would be necessary only for inter-depot transfers and consolidating stock positions and movements. However this would involve greater communications cost (i.e. between the outlets and the warehouse). So one can see that ddp offers a very wide range of possible solutions. Its general principle is that it allows an organisation to map its data and processing capabilities on to its structure.

10

Telecommunications

A significant part of the developments in microelectronics has been fostered by the needs of the telecommunications industry (indeed, the original invention of the transistor was made in a Bell laboratory). Two things in particular have driven these developments — the needs of the military for smallness, reliability and portability — and the requirements of the space program and satellite-based communications systems, where obviously small size and reliability are at a premium.

For many years the telecommunications equipment manufacturers have been major customers of the electronics industry — and more recently of the semiconductor industry. However, three major developments in the way the telecommunications systems of the world work will radically change the relationships between the telephone companies (and their suppliers), the present computer companies, and the semiconductor companies. These are:

> the movement from analogue to digital transmission switching systems;

> the adoption of new, high capacity transmission technologies, such as fibre-optic cable and microwave satellite links;

> the development of new transmission protocols to improve the utilisation of the network.

Before discussing each of these developments in more detail, it will be useful to review the basic characteristics of any telecommunications network.

A Telecommunications Network

It is perhaps important to realise and appreciate the scale and complexity of the world's telecommunications networks — networks which enable virtually all of the subscribers to the telephone system in the industrialised countries to dial a call directly to any other subscriber worldwide, with a very high probability of success! In many ways the world's telecommunications networks are the most complex creations of industrialised society, both technically and organisationally. A simplified diagram of such a network is illustrated in Figure 10.1.

The subscriber is connected over what is termed a *local loop* to a local telephone exchange; a local area call will in general go no further than this and be routed to the called subscriber. A business subscriber will almost certainly have his own 'in-house' local exchange, a Private Automatic Branch Exchange (PABX), which performs the functions of a local exchange for calls within

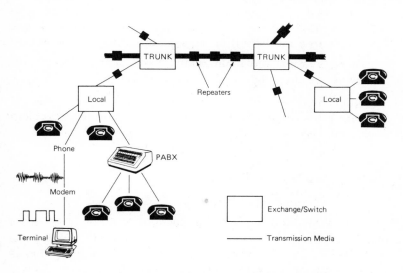

Figure 10.1 *Components of a Telecom Network*

a building or site, and will be linked to a local exchange for calls over the public network. A long-distance call will be routed via one or more trunk telephone exchanges to the local exchange of the called subscriber. To provide some sort of scale for Figure 10.1, the British Telecom network is the third largest in the world: 16 million local loops, 6000 local exchanges, 400 trunk exchanges and 2 million kilometres of trunk lines.

Some of the basic problems seen to some degree or other in the present telecommunications networks are:

> the transmission quality is often poor, which is not of great significance for voice traffic, but is so for computer-related traffic;

> the speed at which the switches in the network operate (i.e. the exchanges) is very slow;

> the utilisation of the physical network is very low, since for voice traffic a subscriber rents an exclusive circuit, but doesn't use its full capacity;

> the switching logic in the exchanges is fixed, so that making changes to the network is difficult;

> there is no 'intelligence' in the network at the exchanges, so that provision of new facilities is difficult, and has previously meant setting up separate, duplicate networks;

> the operating and maintenance costs of the present network are very high, and increasing over time.

Each of the three new developments mentioned in the introduction to this chapter addresses one or more of these problem areas, and we will now discuss each in turn.

Changing to a Digital Network

Methods of Transmitting Information

Since their invention, telephone systems have worked on an analogue basis. In such a system the information to be transmitted — be it voice, data, video, text, facsimile — is represented as an electrical signal varying in frequency and amplitude. As it is transmitted through the telephone network, the signal slowly decreases in strength, so that it has to be passed through several repeaters, each of which amplifies the signal. Unfortunately, any 'noise' that enters the system will also be amplified along with the original message. The result is that the message received at the end of the transmission is usually not the same as that which started out, in as much as there may well be a relatively large amout of noise present, compared with the original signal.

For voice communication this is not too serious, since it is usually possible to infer the content of the message, even if the transmission is subject to considerable noise. However, in the case of other messages, noise is potentially a much more serious problem, and extensive error checking has to be incorporated into such transmission systems in order to try and negate the impact of noise. On an analogue network, any binary-coded information to be transmitted over the network has first to pass through a *modem* (a modulator/demodulator), which modulates a reference frequency to represent the binary pulses. This modulated frequency signal is then transmitted over the network, and at the receiving end another modem converts the modulated frequency back into binary pulses.

In a digital transmission system, the amplitude of the voice signal is sampled at (very) frequent intervals and measured as a numerical quantity (Figure 10.2). Similarly, other media for transmission such as text or images can also be coded numerically, and any messages emanating directly from a computer system will in general already be in binary-coded form. Thus in a digital system all the material being transmitted is numeric, and can be expressed as a sequence of binary digits. At each repeater along the line all the system has to be able to do is to recognise whether a pulse was present (a binary one) or not (a zero), and retransmit it. In the case of digital transmission, the process at each repeater is not to amplify the signal, but simply to re-create an exact binary one or zero. Extra digits are added to each message as it is transmitted, enabling a check to be made at the receiving end to ensure that no information has been lost in the process. Thus in such a system messages may be transmitted with total accuracy, and possibly with total security. Encryption (scrambling) of the analogue signal is difficult, but with a digital signal it is much easier, and if controlled by a computer is potentially foolproof. As the computer manufacturers were quick to spot, computers would be able to communicate directly with such a system, but much more important, they spent all their

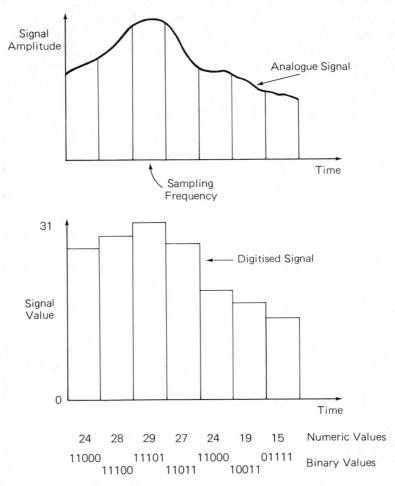

Figure 10.2 *Converting Analogue Signals to Digital Form*

time processing binary-coded information, and thus could provide the necessary intelligence in such a system.

Developments in Exchanges

Basically the provision of some intelligence in the telephone network involves the design and installation of computer-controlled telephone exchanges. Many of the benefits of digital PABXs stem from their programmable nature. For example, these can be seen already on many private PABXs within companies. If a called extension is engaged, the exchange can be instructued to call back when it is free; if a subscriber is temporarily in another location, the exchange can be instructed to transfer all calls to another extension automatically; answering and message handling facilities can be provided automatically at the exchange, rather than at each individual extension; facilities such as abbreviated dialling and repeat dialling can be easily implemented. Any form of binary-coded information can be handled, and full accounting can be provided on an individual or group extension basis. When the public network is converted to digital operation, then facilities such as these can be made available to all subscribers.

The real benefits from such a system will become generally apparent only when the public switched network goes digital, and the main trunk and local exchanges are replaced. Virtually all of the telephone companies in the world (usually referred to as the PTTs — Post, Telephone, Telegraph) plan to convert their systems to digital operation before the end of this century, and the traditional telephone equipment suppliers are all launching their rival systems. However, the vast majority of the telephone exchanges in the UK are based on an electro-mechanical switching technology, known as Strowger, after its inventor. Strowger was an undertaker in the American Mid West at the end of the last centry, whose business was located in a small town. The wife of his major competitor operated the local telephone exchange. Strowger believed that he was losing business because of this, and thus designed and built an automatic exchange. Strowger equipment provided the basis of the world's telephone system for the first half of the 20th century.

Another electro-mechanical system, known as Crossbar, was developed and installed in some European countries since the Second World War, but only in a few exchanges in the UK. Development efforts in the 1950s and 60s were focused on analogue electronic exchanges, but with little practical success. Development took much longer than anticipated, and eventually these analogue electronic exchanges (know as TXE) were overtaken by the developments in digital switching and transmission. Because of the long-standing efforts on analogue exchanges, the UK development effort missed out on the initial move to digital. This was partly due to the situation in the UK in which several companies manufactured equipment to Post Office (as it was then) specifications, and there were allegations that the Post Office specifications were unduly strict and idiosyncratic, thus slowing development and reducing export opportunity.

As a consequence British Telecom is beginning only now to change over to a digital network, and has recently been forced to order more technically obsolescent TXE4 analogue exchanges to replace worn out Strowger equipment. The UK's digital telephone exchange for the public network is known as *System X* (it had its public unveiling in 1980), and is a joint effort by GEC, STC and Plessey, to specifications by British Telecom. In business terms, System X will be 30% cheaper to purchase, 50% cheaper to operate, and 100 times more reliable than its electromechanical predecessors. Computer-controlled telephone exchanges bring many advantages apart from that of accuracy of transmission. The most important are the sheer speed of switching, and the intelligence in switching, enabling advantage to be taken of new transmission protocols, and perhaps above all, much more efficient use of the trunk links between exchanges.

The UK Digital Network

British Telecom has recently announced its plans to begin transferring the UK trunk network to digital operation. A contract has been placed with GEC and STC for equipment for 180 miles of trunk links, making each link capable of transmitting 140 million bits of information per second. British Telecom has not as yet given any date by which the entire trunk network will be digital, but it has said that it will not order any further analogue exchange equipment after 1984. One of the problems of the PTTs in the industrialised countries is that they already have a large complex analogue network in place, and to change over to digital operation will take enormous physical and financial resources, yet the pressure from customers for an improved service is growing inexorably, so that plans for a gradual changeover lasting until the end of the century are under constant pressure. As a reaction to this pressure, in June 1981 British Telecom announced that it would implement a digital, high-capacity network within the City of London area, for a select band of large customers, to provide many of the facilities described in this chapter. In part this plan was also a reaction to potential third-party competition, following on from the relaxation in its monopoly.

In practice, the introduction of digital transmission over the network and of the System X exchanges has to proceed roughly in parallel — ultimately one is of little benefit without the other. British Telecom is at present converting much of the trunk network to digital operation, for example, the terrestrial microwave links carry digital signals (described in the next part of this chapter). Currently, these digital signals are converted to analogue immediately prior to entering the trunk exchange. Slowly the trunk exchanges will be converted to System X, followed by the local exchanges in the major conurbations. British Telecom, not unnaturally, wants to bring the advantages of the digital network first to business customers, who are in a position to make use of it and are willing to pay for additional services. By 1986 British Telecom intend to have the 30 major urban centres on the digital network, and by 1990 75% of customers will have been converted to digital operation.

Transmission Equipment

At the same time as the switch from analogue to digital operation is being planned, there have been major developments in physical communications equipment — in particular, satellites, fibre-optics and the use of the infra-red part of the spectrum. The basic transmission technologies available are as follows:

> 'twisted-pair' copper cable, which has been and still is the basic building block of most telecommunications networks; almost all local loops (i.e. your telephone's connection to the local exchange) use this technology, and unless quite sophisticated transmission protocols are used, its transmission capacity is very low; however, it is a cheap, commonly-available, well-understood technology.

> co-axial cable has a much higher capacity than twisted-pair, and is already widely used for video signals (for example, the cable from your TV aerial to the set); again, it is a cheap, available, well-understood technology.

fibre-optic cable is a very fine strand of highly reflective glass down which it is possible to send pulses of light at very high speed, represnting binary-coded information; it has the potential for very high capacity at low cost; however, it is a relatively new and rather esoteric technology.

infra-red applications have at present been confined to remote-control units for domestic television and hi-fi equipment, but it offers the potential for true 'wireless' communication.

microwave transmission requires a clear 'line-of-sight' between the transmission points, and can be subdivided into two categories; *terrestrial*, using land towers (such as the Post Office Tower and similar constructions elsewhere in the country), which provide a large proportion of the trunk-link network capacity in the UK; and *satellite*, which appears certain to become the major technology for cross-continent and cross-ocean transmission in the future.

Of these technologies, twisted-pair, co-axial cable and terrestrial microwave are already heavily used in the world's telecommunications networks; it is the three new technologies, satellite microwave, fibre-optic cable, and infra-red which appear likely to have the greatest future impact, and which we will discuss next at greater length.

Satellites

Satellites are already used in telecommunications, but a large part of the expected increase in capacity and reduction in cost of telecommunications during the next decade will come from a vast expansion in the use of satellites. With the advent of the American Space Shuttle, the payload that can economically be put into space will be dramatically increased, relative to the cost of using rockets. In a satellite-based communications system the size of the Earth-based unit bears an inverse relationship to that of the satellite-based unit (i.e. the bigger and more powerful the satellite, the smaller and less powerful is the required earth unit). With the size of satellite-based systems that it will become possible to install with the Shuttle, the arrival of Captain Kirk's communicator (as featured in the Startrek TV series) comes much closer to reality!

Satellites used for telecommunications (and direct broadcast-ing purposes) have to be placed in a geo-stationary orbit, 23,000 miles above the equator. At this point, the forces tending to send the satellite out into space and the gravitational pull of the Earth exactly match, so that the satellite remains stationary over a single point on Earth. Typically implementation of a satellite-based service involves the construction of three satellites — two are placed into orbit very close together, one of which is opera-tional, providing a service, the other is available on immediate standby should the first break down. The third is kept as a spare on the ground. Until the end of 1981, only NASA had the ability to place satellites into geo-stationary orbit, but now the Euro-pean Space Research Agency also has the capability. Rocket launches cost a minimim of £10 million, with no guarantee of success, yet there is a waiting list stretching four years ahead. When the Space Shuttle is used for this purpose, very large satel-lites will be constructed in near-Earth orbit from components transported in the Shuttle, and then boosted into geo-stationary orbit for operational purposes. (Satellites constructed in this way will be much larger than anything that could be boosted up directly from Earth.)

Fibre-Optic Cable

Whilst satellite-based systems will provide a large, relatively cheap increase in telecommunications capability (particularly using the new transmission protocols and systems discussed later) for long-distance intercontinental traffic, increases in local traffic (and intra-organisational traffic) are likely to be catered for by increased use of terrestrial microwave systems, and by the introduction of optical fibre systems. Optical fibres are finely drawn fibres of special highly-reflective glass (about 1/10th of a millimetre in diameter) through which pulses of light (represent-ing binary data) travel, initiated by a semiconductor laser, controlled by microprocessor. Optical fibres transmit data at the speed of light, and have immense capacity.

One telephone conversation can be represented by a stream of 64K bits per second. In practice several such conversations would share the same fibre (in technical jargon, multiplexed), and using the current fibres up to 120 conversations could be transmitted at a rate of 8 million bits per second along the fibre.

(New experimental fibres developed by British Telecom can accommodate 2000 circuits.) Plans have already been announced by AT&T to lay a fibre-optic cable across the Atlantic (and one to Hawaii from the West Coast of the USA), to come into service in 1988. The specification calls for this link to be able to handle up to 36,000 simultaneous telephone conversations. If this appears slightly odd, having just discussed the advantages of satellites for intercontinental communications, the reason is that for security purposes a rough balance is maintained between the satellite and cable capacity for transatlantic communications traffic.

The signal loss in using optical fibres is much less than in a conventional system using twisted-pair copper cable. In the current cable systems amplifiers are required every one or two kilometres — in present fibre-optic systems they are needed only every 8–10 kilometres. British Telecom recently announced the development of a fibre operating at 140 million bits per second and a distance between boosters of 102 kilometres. This glass is so transparent that if you could build a window 12 miles thick it would still be as transparent as the average window pane. The implications for this in the trunk network are enormous, since in large cities most exchanges are less than 10 kilometres apart, so that amplifiers would be redundant — the cost-benefits from reduced capital expenditure, lower operating expense, increased capacity and better use of existing trunking are clearly considerable. Such fibre links, together with digital exchanges, form the basis for the future public telecommunications network.

It is just now that optical fibres are emerging from the development laboratories. British Telecom have installed experimental lengths and have placed orders for optical fibres to handle all traffic on 450 kilometres of routes in the early 1980s. The great advantage from the point of view of British Telecom is that the fibres fit in existing cable conduits between exchanges. Thus considerable expansion of the system is possible by replacing hundreds of copper cables by an optical fibre in an existing conduit. Optical fibres are flexible — it is possible to bend them, and even to tie a loose knot in them! As the information is transmitted along an optical fibre as pulses of light, rather than pulses of electricity as in a copper cable, 'cross-talk' (one transmission interfering with those on the adjacent cables) is no longer a problem. The delay in widespread adoption of fibre-optic cable in the public network was caused by problems at a rather mundane level: what happens when it is accidentally damaged or broken and has to be repaired on site? For copper cable this is a tedious but relatively straightforward job. In the case of fibre-optic cable the two ends of the cable have to be aligned with an accuracy of about 0.0001 of an inch, then fused together. It was essential that equipment was made available to maintenance engineers, to enable them to carry out the process, prior to fibre-optics coming into general use.

Infra-red

Whilst satellites and optical fibres will handle communications over longer distances, within an individual room communications and control may be undertaken by semiconductor devices using the infra-red part of the spectrum, introducing a new era of wireless. Infra-red behaves similarly to visible light: it will travel through windows, but not walls; it can be directed or pointed like a torch, or diffused throughout a room. The initial development of these devices (which are very simple by the standards of the semiconductor industry) was for the remote-control units in televisions. At present they are rapidly spreading into the toy market, and TI has recently announced an infra-red remote-control chip set for the toy manufacturers, to go with its processor chips.

Another potential application area is where communication is required with people whose precise location within a room or building is not known. For example, some theatres in London now offer a service for the hard-of-hearing whereby they can rent a set of headphones which receive a broadcast of the play, which is diffused throughout the theatre on infra-red. The advantages are that is is not necessary to run a cable to every seat as would be required with conventional headphones, and the broadcast is restricted to the body of the theatre, which would not be the case with a radio broadcast.

However, the big application, as seen by the computer and communications companies, is as the general communications medium for the office and factory floor. An infra-red diffuser in each room, connected to a central computer (or more likely to a computer-controlled digital PABX), will provide a common shared communications path for the myriad 'smart' office products which are now being introduced — word-processors, intelligent copiers, cordless telephones, data capture terminals, and so on. This will remove the need to install communications cables to each device, so that all units become portable from office to office, and the telephone will become personalised and pocket-sized — and will be extremely difficult to get away from.

An alternative approach to the same problem (i.e. handling communications between many different devices in an office environment through some common carrier system) is provided by the development of local area networks, which are discussed below.

Transmission Protocols

Transmission protocols describe the fundamental set of rules or procedures by which messages are sent over the network. Traditionally the physically network — the cables, exchanges, etc. — and the logical network — the protocols — have not been clearly distinguished from each other, largely because any one physical network implemented only one logical network (i.e. there was a one-to-one correspondence between the two). Now, however, the trend is to maximise the use of the physical facilities by implementing several logical networks (often in hierarchical levels) on a single physical network. Some of the developments in the form of logical networks, and especially the generic group known as *value-added networks* (VANs) are discussed later in this chapter. The sophistication of the protocol used can have a significant impact both on the utilisation of the physical network, and on the services provided to the user. The basic transmission protocols can be roughly categorised as follows:

> *switched voice-grade circuits* (i.e. the public voice network) are the standard network circuits; if used for non-voice traffic by means of a modem, they can only operate at very slow speed (typically 300/2400 baud, or bits per second, bps).
>
> *dedicated high-speed circuits* are leased from the PTTs between fixed points; it is over circuits such as these that large organisations have set up their own private telecommunications networks, to a certain degree independent of the public network.

message-switching networks are based on leased lines from the PTTs and are application-dependent; they are private message-handling systems set up for a specific purpose — for example, the SWIFT (Society for Worldwide International Financial Transactions) network is an international banking system for funds transfer, owned and operated by the participating banks, using leased lines from the PTTs.

packet-switched networks (PSNs) can be thought of as general-purpose message-switching networks, and are an attempt to implement a software solution to the problem of increasing the utilisation of the network (by incorporating intelligence in the exchanges); PSNs are discussed in greater detail below.

integrated communications systems offering sophisticated message handling and transmission features for any type of information media (e.g. voice, text, data, video and facsimile), over a single common network, are the latest arrival in the telecommunications market-place; they also are described in greater detail below.

local area networks are designed to solve the problems of communications within a particular building or site, such that each piece of information handling or processing equipment can communicate via a common network; these too are discussed in further detail.

Packet-Switched Networks (PSNs)

On the present public-switched network, a caller essentially rents a two-way circuit for the entire duration of the call, yet for a high proportion of this time the circuit is in fact idle (especially if the material being transmitted is data rather than voice). One solution for improving the efficient use of this system is for the user to create his own intelligent network using computers to assemble the messages and control their routing over lines leased from British Telecom (many large UK companies presently operate systems such as this, which are basically particular message-handling systems). However, such private systems, though providing a better service at lower cost to the organisation, as compared with what would be available over the public network, do present major technical problems (since the user is now responsible for the switching protocols over his network), and requires a high level of competence in technical management.

A further development of this technique is packet-switching systems, in which messages to be transmitted are divided into small packets of a standard size and format, transmitted through the network at high speed, then reassembled at point of receipt. Users require what is known as a PAD (Packet Assembler Disassembler) to access the system. Such systems give much better utilisation of the network, are much more flexible, and from the point of view of the user, are much cheaper. (A message which cost £4.32 using the conventional system, cost only 16p using the French packet-switched TRANSPAC system.) Charges to the user of a PSN are based on the length of the message sent, and not on the distance between the sender and recipient as is the usual case for sending messages over the public network.

Packet-switched systems over the public network either exist now or will do so in most industrialised countries within a very few years, and will be the standard in the 1980s. British Telecom's Packet-Switched System (PSS), now known as Switchstream 1, was launched as a generally available service in 1981, but ran into some early problems largely as a result of demand exceeding available capacity. On the present analogue network the PSS has to be set up as another network with its own controllers and switches in existing exchanges. In Europe, packet-switching systems are being offered by the PTTs themselves, but in America this market was originally opened up by the creation of Arpanet, a predominantly university and research laboratory-based system. The two major American systems now available, Telenet and Tymnet, lease trunk lines and satellite channels from the network providers and can be accessed from most European countries. A standard protocol — known as X25 — for sending messages over such packet-switched networks has been agreed internationally, and it is relevant to note that almost all computer manufacturers have software available enabling their machines to communicate over networks operating under this protocol.

Local Area Networks (LANs)

As a consequence of the developments in distributed data processing and telecommunications described above, many organisations have more than one source of computing-related resources and have a variety of electronic, binary-encoded information to transmit. In general, it is important to be able to access and use these resources from a single point, over a single network. This is a recognition that of the information generated in an organisation (increasingly in an electronic form) more than 50% is used internally.

One method of enabling a single terminal to access a number of computing resources is to connect both terminals and computers through a digital telephone exchange, capable of automatically switching any terminal to any computer. There are several applications where such a system has been successfully applied and in which it is perfectly appropriate. However, as a general solution it is inadequate since at some point exchanges of this type suffer from the same problems as do all exchanges — in this instance, contention for a limited number of ports. For example, a potential user of a particular resource may find it unavailable, as there is no free connection available. Such exchanges make fixed connections between the user and the particular resource; whilst the link is being employed by one user no other user can access it, yet the capacity of the link is rarely being fully used by the lucky user who has been connected.

Thus there has been widespread development of alternative ways of handling the problem, though to date there have been more product announcements than successful operational systems at user sites. The generic label for the proposed solutions is *local area networks.* Such networks are seen as being within building or site location communication systems — communications over longer distances would employ some of the systems described earlier in this chapter. The orientation behind the development of these networks is towards providing a linking mechanism between different types of device, rather than simply linking lots of terminals to one or more computers. Thus the devices on the network have functional status — work stations, file-servers, processors, various forms of communications interfaces, printers, and so on. There will be a standard interface for each device to communicate over the network, and the addition of new devices or removal of existing ones should be a straightforward matter. As we shall discuss in Part V, the development of successful office systems would appear to be heavily dependent on some fairly standard local network capability. Local networks can be directly linked to other networks of the same type or can communicate with existing computer network protocols or with PSNs. Figure 10.3 shows the functional units in a local area network.

One solution that does find some application is to make use of the packet-switching protocols described earlier in a local area mode, rather than over the public-switched network. As already described, the essence of a PSN is to form a notional virtual circuit for the user, over which his message is transmitted as a multiple set of packets. Implementation of a packet-switched local network is possible (most minicomputer vendors support the internationally agreed X25 protocol), and it would provide a more efficient usage of the physical network than simple point-to-point circuit connections. However, PSNs in this context do have several disadvantages, largely because in their design specification they were targetted at long-distance traffic, using existing low band-width cabling, and employing sophisticated error checking and correction features to cope with the expected high error rates using the public network. The problem in a local area environment is that the user doesn't want to bear all of this overhead, but requires a high-capacity simple interface system.

The designers of the local networks concentrated on increasing the available band-width (i.e. the transmission capacity of the link), primarily by using alternative transmission technologies, particularly co-axial or fibre-optic cable. At present there are two basic types of system competing for the local network business, each of which has several local variants. They are, either some form of 'ring' concept; or some form of linear 'bus', analogous to the data highways internal to some computer systems. The ring concept originated in the UK at Cambridge University (and hence is usually known as the *Cambridge Ring*), though present development of a commercial system is being undertaken by several software suppliers and system houses. The ring, as its name suggests, is a complete circuit loop formed from standard twisted-pair cable — messages are broken into digital packets and sent in a high-speed burst. Each device attached to the ring is able to gain control of the ring in order to send a high-speed packet to any other device. Intelligence in the ring detects that a particular device has control and is transmitting, and denies control to any other device wishing to transmit, forming a queue of such requests if necessary.

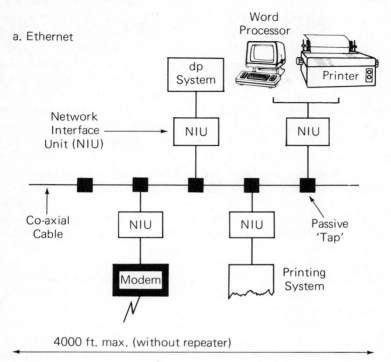

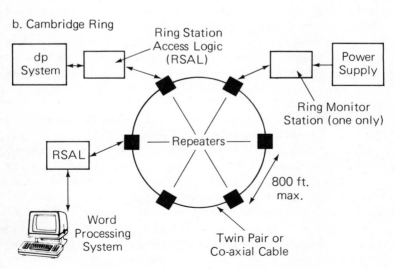

Figure 10.3 *Examples of Local Area Networks*

Probably the most widely known of the bus-type local network systems (and probably the most widely known local network system in general), is a system known as *Ethernet*, originally developed by Xerox in the early 1970s, but appearing as a commercially available system only in 1980. Ethernet uses a single co-axial cable, to which its devices are attached. Essentially it is a broadcast system, with no intelligence in the network itself. The device wishing to transmit a message looks at the network to see if a message is already on it. If so, it waits a very short random length of time before trying again; if not, then it broadcasts its message over the network. The normal mode for all of the devices on the network is in receiving (i.e. listening) mode, and all have a unique network identifier by which to recognise messages intended for them. The identifier of the device for whom the message is intended is included as a part of the message; as in the ring system, messages are transmitted in packets at high speed. There is a very small chance that two devices may try to transmit simultaneously. The sending devices are able to detect that such a collision has occurred, and both re-transmit their message after a random small interval of time.

Xerox has formed a business grouping with Digital Equipment Corporation (DEC), the largest minicomputer manufacturer in the world, and Intel, one of the major chip manufacturers, to exploit Ethernet commercially. Xerox also makes a licence for the Ethernet protocol to be made freely available to other organisations interested in its use, in an attempt to foster its development as a *de facto* standard for local networks. Ethernet is what is known as a *baseband* system which has only a limited capability to handle voice traffic, and too low a band-width to cope with video transmissions.

Several other manufacturers have built higher-level standards on Ethernet in an attempt to counter some of its deficiencies. For example, in mid-1981, Wang announced its Wangnet as a super-set of Ethernet, able to handle both voice and video, as well as text and data. Each of the different types of transmission has its own channel over the communications link (usually twin co-axial cables), and this type of system is known as *broadband*.

Technically, the difference between baseband and broadband is that in the case of the former, the frequency spectrum varies from an absolute origin in only a single direction, thus limiting its capability. In the latter case, the variation is in both directions from a relative origin, giving greater capacity.

Integrated Communication Systems

It is now apparent that as a consequence of these developments, the artificial distinction between data processing and telecommunications (imposed by legislation rather than any coherent view of the market or needs of the user) will disappear in the information processing revolution of the 1980s. Integrated, intelligent networks capable of handling voice, data, video, text and facsimile will become the norm. In the USA the communications industry is being deregulated, which is bringing companies such as IBM and Xerox into the field; and the converse is also true: the recent removal of regulations restricting the business ventures of AT&T will allow them to enter the data processing market.

As an example of the likely characteristics of such integrated communications systems, let us briefly examine the system offered by Satellite Business Systems. SBS offers an integrated system designed to handle the *total* communication requirements of a large organisation, both internally and to other SBS users (initially only in the USA). The user of such a system will merely have to indicate the recipient of the message (or recipients, as such systems will circulate the same message to multiple recipients). The system itself will cope with routing the message, coding, encryption, data checking, and so on, and will assume responsibility for its accurate delivery whatever the originating media type. Each user will have a direct computer-controlled link to a local Earth station, which will communicate with a satellite. The system is based on existing operational technology and became available early in 1981 in the USA.

SBS Inc. was formed as a joint venture by three companies: Comstat (a satellite technology company), Ataena Life (one of the largest American insurance companies) and IBM. As the data transmission speed with the SBS network is comparable with the internal transmission speeds within a computer, there is the slightly unnerving prospect of all IBM computers and PABXs being part of a worldwide network able to communicate automatically with each other (in an unbreakable code!). The difficulties in achieving what SBS now has operational should not be underestimated. Both Xerox (with XTEN) and AT&T (with ACS) announced some years ago that they too would offer similar systems. Xerox dropped out of the race, and AT&T is still working on it. Though to be fair, AT&T's Advanced Communication System is in many ways more revolutionary than SBS, as it is aiming to connect a wide range of equipment from different manufacturers.

Viewdata/Videotext

Viewdata, or *videotext* as it is now more widely known, is a technique for storing, accessing and retrieving data from a remote computer using an adapted television set and a telecommunications link (usually the public telephone network).

Information is stored on a viewdata computer as a series of *frames* or *pages*. Each page consists of both content and routing information to the next pages. The information is stored as a massive *'tree-structure'* with an initial menu page leading to other menu (routing) pages until the page containing the required information is reached. The beauty of the system is the simplicity of use and that it is totally content-free (i.e. any kind of information can be stored and retrieved by this system). The disadvantage is that it can be a rather cumbersome process finding the required information. Potentially the database has unlimited capacity and any page can be requested directly if the correct page number is known.

To access such a system one needs either a viewdata television or a special adaptor. The adaptor contains a modem to decode the information into binary form, some memory to store the page of information, character generators and refresh circuits. A key pad is also needed, which can either be solely numeric with two extra keys, or a full alphanumeric pad. The viewdata computer has to be dialled up (automatically or manually) and a password given to identify the user. Any page of data can now be requested, and special response frames allow the user to send 'transaction' data back to the viewdata computer. This is illustrated in Figure 10.4.

Viewdata must not be confused with *Teletext*, a system for transmitting 'frames' of information using normal television broadcasting techniques. This system was initially developed by British television engineers as a way of putting subtitles on the screen for deaf viewers. Teletext is a one-way system with a limited number of frames (a few hundred) being transmitted in sequence and the receiver has to wait for the required page to come around.

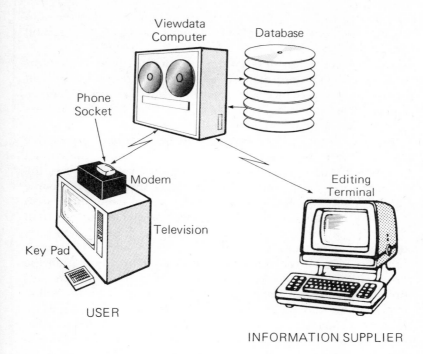

Figure 10.4 *Viewdata*

Prestel

The first application of viewdata was by the British Post Office in setting up the Prestel service. This was the first public system, launched in 1979. The plan for its development was grandiose to say the least, and totally unrealistic on two counts. First, the operational plan required the installation of viewdata computers (time-shared) on an unprecedented scale, and the computer (made by GEC) together with the local exchanges would have probably been unable to cope with the load. But the second and more serious problem was a total misreading of the market. The idea of Prestel was to provide a domestic viewdata service to the public at large. Unfortunately the public still has a view that 'information is free'. The cost of the system was high both in the capital cost of sets and in the operating cost. At present the public has little use for a straightforward information retrieval system; instead a system that more closely met a real need would be one that enables them to make transactions, after getting availability information.

However, the business community found these limitations less of a problem. Information was not regarded as free, speed of access and convenience was appealing, and the cost not such a barrier. Consequently Prestel has grown up to be a business information system and a successful one at that.

Before Prestel can become a widespread public system two things have to happen. First, costs have to fall — the cost of receivers has to come down to a marginal level (relative to the existing cost of a TV set), so that a large critical mass of people can have access to the system; and the cost of using the system will have to drop. Second, British Telecom will have to get out of the information warehousing business as Prestel was first conceived, and back to running networks which is its proper function. The current major technical limitation is that all the information in the Prestel database is fed in through terminals operated by the *information suppliers*. Thus often the information suppliers are having to reproduce information already held in their own computer systems. To overcome this and other limitations a new technique called *gateway* has been developed (ironically by the Germans who bought the original Prestel system and then sold this revised version back to us). Using the gateway approach (see Figure 10.5) the viewdata computer no longer needs to store the information locally, but accesses the information suppliers' computer directly. This means there is only one copy of the data and that transactions can be sent directly back to the information supplier. The Prestel system can then become solely a network that allows the public to communicate with the computer systems of a large number of organisations.

Currently Prestel is used for financial information (the Stock Exchange, dataSTREAM, Fintel): timetables and schedules; news, sport and weather reports; government statistics; holiday information; mail order, and games. These are all services with public access, but within the public system it is possible to set up a *closed user group* with access limited to authorised people. This could be used for a customer information system; for example, GKN unsuccessfully tried to set up their stock availability information (this should become quite feasible using the gateway approach). Several organisations have set up sales information

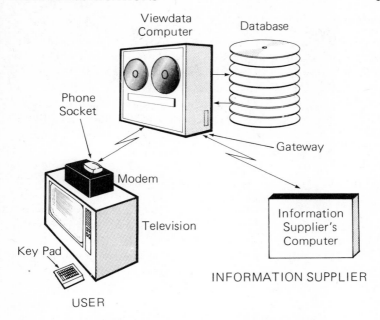

Figure 10.5 *Viewdata — Gateway Approach*

systems where details of sales are passed to and from regional sales representatives using Prestel. But possibly the biggest potential application of this approach could be the clearing banks. Using the gateway technique, customers could get statements, authorise payments, standing orders, in fact handle most of the transactions normally done through a local branch.

Private Viewdata System

Prestel is but one application of viewdata — the major successes have been with private organisations installing their own systems for their members or customers. Viewdata software has been implemented on many different computers, indeed the technique came of age when IBM announced their own version on an IBM Series 1. Possibly the most successful company in this field has been Rediffusion Computers (previously Redifon) who claim to have sold more viewdata systems than anyone else. One of their biggest systems was sold to Thomson Holidays to provide an information and booking service for travel.

Electronic Banking

It is in the world of electronic banking that telecommunication provides a principal agent of change. Banks have for many years transferred monies 'over the wire' to other banks. Today it is estimated that the volume of interbank transaction in electronic form may be as high as $100 billion per day. Indeed, the banks have set up their own telecommunications network, SWIFT (Society for the Worldwide Interchange of Financial Trans-actions), to facilitate this process. Using this network it is possible for, say, a Hong Kong bank to clear a cheque on a Californian bank within a few minutes. The SWIFT network consists of a series of lines leased from national PTTs, with small computers acting as switching modes. All messages on the network are routed to two large Burroughs computers in Brussels where they are recorded for security purposes, then switched on to their destination. This is an example of a *message switching system* and it may be classed as a *value-added network*.

However, the major banking development being spurred by telecommunications is in retail banking. The costs of handling the vast number of small transactions that make up today's retail banking have been kept high by the large number of people involved in processing them. It is not that banks have failed to invest in automation, quite the reverse, but the interface between the customer's transaction and the bank has still needed human intervention.

The UK clearing banks (like other banks the world over) are presently investing heavily in automated bank teller machines — not just cash dispensers, but machines which can fulfil several other functions, such as ordering cheque books, requesting statements, and so on. From the bank's point of view such machines are wonderful — they use the pavement as a banking hall, kindly provided by the local authority, and externalise their labour costs, much as self-service has done in the rest of the retail business. And there is a lot of evidence that customers prefer using the machines too; certainly if you want to withdraw only a small sum it saves you from the rather condescending gaze of the teller.

The next developments in this area will be to move automated retail banking into other locations than bank walls — factories, offices, shops, railway stations, and so on — basically almost anywhere that a free banking hall is available. The final logic of this process is to move much more of this right back to the home and to enable the customer to initiate financial and administrative banking transactions from his home, probably using Prestel. This is still some years away at present, but some banks in Germany are already conducting experiments along these lines.

The development of credit card payment systems reduced the volume of transactions that the banks have to handle, but this load has been transferred to the credit card companies themselves. Each transaction generates a paper voucher that has to be individually processed. In the UK the two major organisations handling these cards, Barclaycard and Access, are finding that the possibilities of further growth are being curtailed by the volume of paper.

Recently an experiment was conducted with a series of garages in the Norwich area, where they were given a device which could automatically 'read' a Barclaycard, and transmit the transaction information directly to Barclaycard's central Northampton computer site. The results of this successful experiment showed, not surprisingly, that the major benefits were to Barclaycard themselves, rather than to the garages or the oil companies; so that if this type of application were to become widespread, then the credit card companies would have to provide the bulk of the investment. This type of system is known as an *electronic financial transaction system*, where the transfer of funds is initiated by a separate process (i.e. paying Barclaycard).

The device which read the Barclaycard in the garage is an example of a *point of sale* (POS) terminal. Similar terminals, which can potentially read credit cards, are being installed by the large retailing organisations for applications such as stock control and cash management. As yet these haven't been connected directly to the credit card company's computer, but rather just work with the retailer's own system. With the development of more economic telecommunication networks (e.g. packet switching) it will become increasingly attractive to connect directly.

In the UK the direction of the credit card companies is dictated by the major banks, as they are the owners. In the US the situation is rather different. Visa and Mastercard have co-operative arrangements with the banks as they offer complementary services. US banks have issued one or other (sometimes both) of these credit cards to 120 million customers. Recently these two organisations have been trying to convince the banks to issue a new type of card called a *direct debit card*. This type of card initiates a direct *electronic funds transfer* (EFT) between two parties. This in effect puts the two organisations in competition.

The US situation is complicated by the 1927 McFadden Act which prohibits interstate banking. Traditionally the small banks have welcomed this protection against the financial giants, but in the last eighteen months there has been a rush to join one of the regional *automatic teller machine* (ATM) networks. These networks allow transactions between customers of different banks. All of a sudden the small banks have realised that they can be lighter on their feet than their bigger brothers. The new networks they are installing, using a packet-switched approach, are considerably more economic than the 'obsolete' networks used by Visa and Mastercard. Using the new ATMs a customer will be able to withdraw or deposit cash and make balance enquiries, the new networks will be capable of large-scale retail fund transfers; and within a few years widespread connections to the retailer's point of sale terminal and home banking terminals will be practical.

In America the speeding progress towards electronic banking is also being fuelled by the competition between major financial service institutions. American Express, Merril Lynch and Sears, Roebuck are all developing different services which will compete for the same funds. Citicorp has made a major commitment to worldwide electronic banking in an effort to make the customer and the computers do the work and reduce their processing costs.

In the UK it has often been assumed that because there were only a few countrywide organisations (i.e. the clearing banks) involved with retail banking, it would be easier to set up transfer networks. It is extremely doubtful whether money will ever be displaced by plastic cards. In the UK only 60% of the workforce have bank accounts (compared with 90% in Europe) and cash still accounts for nine out of every ten transactions. Doubtless these proportions will change, but it is likely to be a long time before EFT offers the same convenience as money.

11

Office Products

The third of the industries converging in the information technology area is that of office products. Unlike the computing and telecommunications industries which are relatively homogeneous, office products comprise a wide variety of products and an even wider range of suppliers. We will describe the developments in each of the main product areas, beginning with the most developed, word-processing — the automation of the typing function.

Word-Processing

Word-processing is concerned with electronically processing textual information in a similar fashion to more traditional data processing — that is, entering, manipulating, editing, storing, retrieving, displaying and transmitting such information. The present suppliers in the market have their origins in two distinct and traditionally separate products, typewriters and computers.

The first typewriter intended for commercial sale was produced in 1873 — with that well-known feature the QWERTY keyboard, and although detailed design improvements were made over the years there was no fundamental change in design for almost 90 years, until 1961 when IBM introduced its electric golf-ball typewriter (though even this type of mechanism had first been invented in 1908). From then on, innovation has been much more rapid, primarily orientated towards providing the typewriter with some capability to store and edit textual material.

Electronic Typewriters

The first progress along this route came in 1964 with the introduction of the IBM Magnetic Tape Selectric typewriter, followed in 1969 by the Magnetic Card version. In both these machines the material being typed was simultaneously stored on some magnetic media — either tape or cards. The typewriter was equipped with a unit which could read a tape or card, and automatically type the contents on the typewriter. Editing of text stored in this way was accomplished by playing back the tape or card; stopping at the appropriate point; adding, deleting or modifying the text; then playing back the rest of the material.

The present stage of this technology is the electronic (or as it is sometimes known, memory) typewriter: a machine with few moving parts; probably employing a 'daisy-wheel' print mechanism (a system invented by a Xerox subsidiary, where the printing elements are the ends of the spokes of a small wheel); some memory capability, probably in the form of semiconductor RAM; and usually with a single-line display that can show the user what has been typed (and can be edited) before committing it to paper. The limitations of such machines lie primarily in the hardware — typically the memory is fairly limited, and the single-line display allows only that line to be edited, making large-scale or global amendments difficult. Such machines are really only suited to correspondence typing and similar relatively short documents. The dividing line between such electronic typewriters and a word-processor is unclear — most electronic typewriters have the capability to be upgraded to form part of what would be universally recognised as a word-processing system.

Word-Processors

At present, the term word-processing seems to be reserved for systems which are screen-based (thus having a separate typing unit) and which have some storage medium which is machine readable and user removable (including magnetic tape and cards, though some form of floppy disk is probably by now the most popular). Whereas electronic typewriters may cost less than £1,000, a word-processing work station will probably cost at least £3,500 (based around a personal computer system), ranging up to £7,000 for more sophisticated systems. The differences between the systems lie in both the type and scale of the hardware used and the capabilities and sophistication of the software.

The major hardware differences lie in three areas: first, the size of screen used, in terms of the number of lines of text displayed. A typical vdu screen used on a computer system displays only a *partial page* (typically somewhere between 20 and 40 lines of text) so that *full-page* displays have been developed for the word-processor market, and may well become the standard in the near future. The second difference is in the form and capacity of the memory available for text. As already mentioned, some form of floppy disk storage has become the norm, using either mini-floppies (5¼ inch diameter, storing about 350K characters), or standard size (8 inch storing 1 million characters).

Third, there is a fundamental distinction between *stand-alone systems*, where the screen, processor, memory and printer are available only to a single user at a time, and *shared-logic systems* in which the user still has an individual screen, but shares the processor, memory and printer with other users. A shared-logic system may be dedicated solely to word-processing, or may be a general purpose computer system, usually a minicomputer system, running word-processing software simultaneously with other data processing applications. Users may be psychologically and administratively much happier with a stand-alone system in the sense that it is their system — all the components are there in front of them, available whenever required.

But for an organisation of any size a shared-logic system has several advantages, although some characteristics fall into both categories. A shared-logic system can be economically more attractive in that the most expensive peripherals (printer and disk) are shared. However, if either of them fails, then none of the users can use the word-processing system, rather than only one user if the disk or printer fail on a stand-alone system. Because the disk is a shared hard disk, the individual user has to rely on others to devise and implement satisfactory archiving and backup procedures. Nor are the disks themselves removable, as are the floppies in a stand-alone system, so that the users lose the (psychological) benefit of being able to look after their disks, containing their material. However, sharing a large hard disk between several users means that it is easier for them to share common material or services. For example, a natural progression from word-processing is into some form of electronic mail service — some means by which people can communicate electronically; this is much easier to implement using a shared-logic system.

Alternative Methods of Operation

From the point of view of software, there is a basic set of functions that any word-processing system must possess. Differences exist in the degree of sophistication in the way these basic capabilities are used and in the extensions in function over and above the basic requirements. The basic functional requirements necessary in word-processing software are that it should be able to accept, edit and amend text; store, retrieve and merge it (with other text); and display it as originally entered or in some user-specified layout. However, there are two very different ways that this software can be implemented.

The first, rather appropriately, is known as *'what you see is what you get'!* Using this type of software, users would typically indicate (amongst other things) how they wanted the final document to be laid out on the page (i.e. would specify such things as left and right margins, page length, left or right justification or both, page numbering information, and so on). Then as text is entered it would automatically be formatted according to these specifications on the page. Similarly, the results of any editing of the text — modification of specific character strings, insertion or deletion of blocks of text, merging and moving blocks of text, and so on — are immediately displayed on the screen in the document's final form. Generally speaking this mode of operation is easiest for first-time users, and those with a non-computing background, to become familiar with.

However, this type of system does place a very heavy load on the processor, in that it is having to manipulate a large block of text instantaneously and is often having to completely re-format the text on the screen (for example, insertion of a new line near the beginning of the document would probably mean that the whole document had to move down one line — insertion of a new word would mean re-formatting a whole paragraph). Thus word-processors of this type are not suited to processing lengthy documents, and in a shared-logic configuration would require a fairly powerful processor in order to provide good response times to the users. Most purpose-designed stand-alone word-processors have software which operates this mode.

The alternative approach is derived from that strand in the development of word-processing which originates from the data processing environment, particularly that of interactive time-sharing systems. The development of on-line, interactive time-shared computer systems during the 1970s soon led to a require-ment for an on-line *editor* to manipulate files of source code stored in character format. The editors developed to meet this need were powerful and capable of handling large volumes of text efficiently. Over time they became powerful, fairly easy-to-use, reliable pieces of software — and as such were developed for most computer systems, especially minicomputers, as they were often one of the primary interfaces between the user and the system. Thus on all minicomputer systems there already existed such a powerful piece of software, and also one typically with a large body of users.

Given that such systems were also multiprogramming sys-tems, then it was logical to add one further program, a *formatter*. This program took as its input a text file with certain formatting commands specifying how the text was to be laid out on the page embedded in it. In this system users are normally working with the original text file as entered and subsequently modified, **not** with the formatted text as in the 'what you see is what you get' system. Indeed, in these systems, users typically don't see the final formatted document until it appears in its final printed version — in the former system they are working with the formatted version all the time.

Thus most users with no previous computer experience find these systems more difficult to work with, though, naturally, people familiar with using time-shared computer systems find few problems. Systems such as these are fairly easy to imple-ment on time-shared, multiprogramming systems, and have the advantage that they place a relatively light load on the processor (since formatting the text, which is the operation that places by far the greatest load on the processor, is generally only under-taken infrequently when the document is finalised). Such systems are perhaps better known as *text-processors*, rather than word-processors, since they are more suited to handling large documents, subject to frequent revision (such as computer software documentation), rather than, say, form letters and mail-shots which tend to be the staple diet of the true word-processor.

Facsimile

Facsimile (usually referred to by the abbreviation 'fax') is a method by which a copy of an image (on paper) is transmitted and reproduced electronically at a remote site, using the telephone network as the connection mechanism. Facsimile was invented in 1842, but widespread commercial services were not available until the 1920s (in America). Until recently fax was seen as a rather specialist service with a limited market (for example, transmitting weather maps, or newspapers). The recent re-emergence of fax is due in part to a perception that reprographics-related products and services are likely to be a significant component of the electronic office.

All fax equipment operates in a similar manner — the page to be transmitted is scanned on a fine grid pattern, and at each point the relative darkness (i.e. between white and black at the extremes) is recorded, transmitted and reproduced at the remote location. The relative intensity can be transmitted as either an analogue signal or as a numeric value in binary form. The system simply scans a document and reproduces the image of it — it has no knowledge or awareness of the content of what is being transmitted. The image on the document being transmitted may well be text or data, but they are not transmitted as characters or numbers but as the appropriate symbol. Thus fax is especially useful in applications where graphics are to be transmitted (for example, engineering drawings), or where a special image has to be reproduced (for example, signature verification).

Fax equipment differs in its scanning density, copy quality and transmission speed and is constructed to meet one of the standards of the Consultative Committee for International Telephone and Telegraph (CCITT) — this is the PTT's international standards committee. Whatever standard the equipment is manufactured to, there is in each category a basic trade-off between scanning density (and thus copy quality), and transmission speed — the higher the density, the longer the scan and thus the transmission time. The CCITT has formulated three standards for fax equipment, namely:

> Group I machines which transmit an A4-sized page in analogue form in 4–6 minutes.

> Group II machines do the same but faster, transmitting an A4 page in 2–3 minutes.

> Group III machines operate digitally and transmit an A4 page in less than 1 minute.

The current market environment is moving rapidly towards the new Group III machines, even though the present boom in fax in North America was sparked off by the availability of cheap Group II equipment.

The existence of the CCITT standards ensures that users can operate with other users employing equipment from different manufacturers, so long as they are within the same Group. British Telecom and the other PTTs have set up a variety of fax services — for example, British Telecom offer their Bureaux Fax service within the UK between several major cities, and the Intelpost service to North America, Europe and the Far East. The major disadvantage of these services is that transmission and reception is possible only at a relatively few centres. The sender has to deliver the document to be transmitted to the nearest centre, and the recipient similarly has to have the copy collected from their nearest centre. Thus in America, ITT and other companies have established public fax service networks in which the customers provide the fax equipment at their locations and ITT provides the network using leased lines — to the user it appears to be organised in a similar fashion to the telex network. A public fax network of this type is a good example of what is known as a Value-Added Network (VAN) service.

Teletex

Teletex is a service offered by the PTTs enabling electronic transmission between Teletex-compatible terminals on a memory-to-memory basis. It is perceived by the PTTs as a way of retaining and developing their telex market and moving into electronic mail services. The document to be transmitted is prepared locally at the memory typewriter or word-processor, and then transferred to the Teletex part of the terminal from where it can be transmitted to another Teletex terminal at speeds of up to 240 characters per second over the public network. Use of word-processors as the originating device means that messages can be created, edited and stored using all of the features of the word-processor prior to transmission.

The Teletex system will also have an interface to the existing worldwide telex network, and it will be possible to send messages between the two systems. An international standard for Teletex should shortly be agreed by CCITT. Several European countries already operate services such as this (notably West Germany and Sweden), and services in North America are scheduled to begin soon. British Telecom intend to launch their Teletex service in 1982, though initially the transmission speed will only be at 120 characters per second. Teletex will be an important component in the development of electronic mail systems, especially for the smaller organisation — for large organisations or those generating a high message volume the transmission speed available on the public Teletex services is likely to be too slow, and they may well use higher speed services implemented on their own private networks or those offered by third parties.

Reprographics

Just as the typewriter has been improved by the incorporation of some form of intelligence, as in a memory typewriter or word-processor, so the other commonplace item of office equipment, the copier, has also benefited from the incorporation of a processing capability. The basic reprographic process in which an image is placed on an electrostatic drum and then transferred to paper many times is well-understood and commonplace. The new versions of the copier, now marketed by most computer companies as well as by Xerox, are based on a laser-printing system. In these systems a laser under computer control 'paints' the image to be reproduced on the printing drum. The image to be printed — text, data, graphics (including signatures, logos etc.) — is presented to the printer in digital electronic form, typically from a host computer system, though most of these printers do have a communications capability, enabling them to accept material sent over a network.

As the image is in digital form it can be processed before reproduction. For example, the image could be reduced or rotated, the type font changed at any point, material from several sources merged onto a single page, and standard forms or layouts stored in the printer's memory can be overlaid at will. Facilities such as this provide a considerable improvement in operational flexibility over conventional copiers, and a consequent increase in throughput. The laser system creating the image works to an extremely high degree of accuracy, so that the quality of the output is very high — almost typewriter quality. The final benefit of these machines is their speed. The conventional printing device when letter quality output is required is the daisy-wheel printer, which provides excellent quality, but at a relatively slow speed (typically 45 or 55 cps). The high-speed printers available on computers don't produce output of an acceptably high quality for typical office applications. The laser-printing systems operate at speeds of 45 pages per minute or more — this would typically give them a 50 to 1 speed advantage over daisy-wheel printers. However, there is one major drawback: cost. At present these systems are priced in the $50,000–$100,000 range, depending upon the options chosen. Thus at present they can only be justified in very high-volume applications, often replacing conventional computer line-printers. However, prices are likely to fall somewhat, and once local area networks become well-developed and the cost can be shared over the network, then laser-printing systems seem certain to become an integral part of electronic office equipment.

It is this convergence of small business systems, distributed computing systems, interactive processing and on-line databases, communicating word-processors, electronic mail, intelligent communicating copiers, computer-controlled message switching systems, and so on, which we believe will form the basis of the revolution in information processing. The convergence of these technologies in information processing has focused attention on the office, an area which will obviously be significantly affected by such developments. The possible scope of office automation varies greatly from commentator to commentator, and it is the purpose of the next section in this book to examine the scope of this concept, and look at some of its potential impacts and consequences.

Part V

OFFICE AUTOMATION

In this part we shall be looking specifically at the effect of information technology on the office environment, and especially at the automation of the office. The term office automation means different things to different people, from distributed data processing to word-processing, from local area networks to management work stations, from microcomputers to complex computer systems. It has also been given a number of other labels such as The Electronic Office, The Paperless Office, and The Office of the Future.

12

What is an Office?

Data Processing and the Office

To put this discussion into perspective, it is worth recalling that the first commercial data processing computer was called LEO, Lyons Electronic Office. The objective of these early dp applications was to increase the productivity of the office. Thus the first stage of office automation began more than twenty years ago. There is much academic discussion about the meaning of the term automation and how it differs from mechanisation, which we will review in the following pages. But these early office applications all had similar characteristics: they all involved a high volume of transactions (to justify the high costs of latter-day computers), which were concerned with well-defined problems (highly structured in computer jargon). These kinds of problems were mainly in the accounting and administrative functions of the organisation. One can look at the development of dp applications in terms of volume and structure, as shown by Figure 12.1.

Traditional dp applications began with those having the characteristics of the top left-hand corner of Figure 12.1, that is, those with high volume and structure. Once these applications were mastered it was possible to develop software which was more complex and could automate less structured procedures (1). These would be systems where the users drive the software down their own chosen paths (i.e. the user is providing the

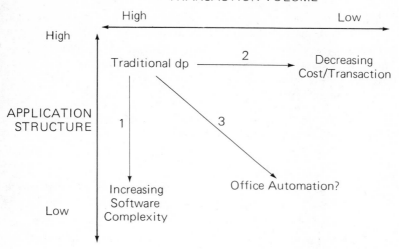

Figure 12.1 *Development in dp Applications*

structure). However, it was still necessary for the volume of transactions to be high to justify the cost of developing this software. A manufacturing information system would be an example of this type of system.

The second movement (2) was a result of the falling cost of hardware, which allowed low volume applications to be developed, as the processing cost per transaction fell. This cut hard into the puritan dp ethic, where idleness was considered a mortal sin and the thought of computer equipment sitting idle for most of the day was not easily accepted. Such systems are characterised by the multitude of small business systems which automate functions traditionally tackled by dp (e.g. financial accounting) for smaller and smaller organisational units.

This leaves only the bottom right-hand corner of Figure 12.1 undeveloped (3). It is a space which has only now become possible to think about occupying. This is because the hardware is now sufficiently cost-effective for low-volume applications to be economic, but the problem of efficient production of complex software is still unresolved. We would argue that the second stage of office automation, which is the subject of this section, belongs fairly and squarely in this segment of Figure 12.1.

Aspects of the Office

Before we can discuss the practicalities of automating an office, it is necessary to have some idea of what an office is. It is an interesting exercise to ask this question of a room full of managers because of the variety of answers that will be given. We list below some different aspects of an office; the order is in no way significant.

Location — the office is a physical place separate from the home, where people go to work.

Meeting Place — the office is a place where people interact for both business and social reasons.

Administrative Centre — the office is the administrative centre of an organisation; alternatively, it may be distributed around a number of small administrative units.

Management Support — the office is that part of an organisation which contains and supports the management functions and personnel.

Facilities and Tasks — the office is a place containing a number of facilities needed in order to perform office tasks.

Functions and Procedures — the office is a collection of functions which can be specified as a set of procedures.

External Relationships — the office is that part of the organisation which handles the communications with other organisations. Very often this involves providing the interface between some external dp system, such as a customer's stock replacement system, and a corresponding internal dp system, in this example the sales order processing system.

System Interfaces — the office provides the interfaces between a number of information systems. This mainly involves collecting data to be put into dp systems, much of which has usually been generated by other dp systems.

From this list it is obvious that the office can fulfil a number of different functions at the same time, dependent upon the type of organisation being considered. There are essentially three different types of office: the clerical office, the administrative office and the management or professional office.

A *Clerical Office* exists where an organisation deals mainly with paper and information; in such an organisation the office becomes the shop floor. In this respect the clerical workers pushing pieces of paper are little different from the blue collar workers with their mechanical tools in a manufacturing organisation. Typical examples of organisations with large 'clerical offices' are insurance companies and banks, where a situation still exists in which an enormous physical volume of pieces of paper is handled.

Administrative offices are usually found in a manufacturing or service organisation, where the office performs an administrative support and control function, making sure that operations run smoothly. The function of this office is not an end in itself, but provides a service to the rest of the organisation.

In contrast, the *Management/Professional Office* is set up to support a number of skilled managers or professionals, often termed 'knowledge workers'. This could be a typical professional firm like accountants or solicitors; or alternatively, it might be the group headquarters of a multi-divisional company. In some instances it might also be called a secretarial office for it will be mainly staffed by secretaries.

Each of these types of office has a different way of working and requires a different type of solution if automation is to progress further.

Information Flows

Another way of looking at the office is in terms of the information flows within an organisation. In Figure 12.2 we depict some of the different personnel, the information that flows between them, and the functions they perform within an organisation. At the bottom level data is captured, classified and stored. This information is filtered and structured by the middle managers and supervisors before being passed to senior management. Senior management will set or modify strategy, and contribute to the formulation of plans and budgets. The information is then passed back down to the middle managers to administer the agreed plans and control operations. Again this gives us a different insight into the office and shows its workings from a different angle. All of these perspectives are important if we are to be successful in 'automating the office'.

Figure 12.2 *Office Functions and Information Flows*

Automation versus Mechanisation

There is much academic debate about the difference between
mechanisation and automation. This may in practice be purely a
semantic argument; however, the case is clearly stated in
Michael Zisman's article in the *Sloan Management Review* (4,
1978). He applies Richard Nolan's stage hypothesis of EDP
growth to office automation. Zisman argues that at the mechanisation stage the office is viewed in terms of the discrete tasks that
are carried out within it, such as typing, filing, message handling, calculating etc. Each of these tasks can be mechanised by the
provision of a suitable tool; for example, word-processor, data
storage and retrieval systems, electronic mail, calculator, etc.
There is a direct parallel with the mechanising of industrial tasks;
each task is viewed in isolation and a solution is found for

increasing local productivity. However the local gain may be more than offset by some negative side effects. For example, if a secretary's typewriter and paper filing system are replaced by a word-processor and electronic data storage system, then this will hopefully increase the productivity of the secretary; but if the manager is unable to use these tools when the secretary is absent then the manager's productivity may be significantly impaired.

In contrast automation looks at the total process as one inter-related system. The system must be analysed to establish the basic functions; methods can then be found to automate these functions, with special care to integrate all the parts. Examples of this kind of automation can be found in the industrial world, particularly amongst the automobile manufacturers. The new Metro production line developed by BL is an example of a situation in which the process has been designed as a whole. This kind of system thinking is not yet prominent amongst office system suppliers. The reason for this lies mainly in the area we have just discussed; namely, that the role and function of the office may not be easily defined and this definition will vary greatly between organisations. However, such a systems-oriented approach will soon become a necessity for the suppliers of office automation systems as organisations begin to integrate their existing electronic office tools, primarily through communications networks to link electronically their word-processors, filing systems, reprographics systems, dp systems, and so on.

The immediate objectives of the organisations in seeking to integrate and develop their automated office systems may be of a general, more imprecise nature — such as improving the effectiveness of large-scale administrative functions, and of co-ordination between decentralised units (for example, in both public and private bureaucracies); or they may be procedures at higher structural levels in the organisation — such as integrated order processing, or manufacturing control. For example, the present state of office automation emphasises technical building blocks aimed at supporting individual, personal activities — such as word-processing, electronic mail and analytic tools. These are tools which appeal to the 'knowledge workers', and support specific tasks that they perform. Yet from an overall organisational point of view the need may be to install systems addressing global functions.

The Nature of Office Work

This leads us to identify a third dimension to the tasks under-taken by the occupants of the office, building on our simple structure illustrated earlier in Figure 12.2. Some degree of structure in the task is retained, whilst our original concept of volume is better expressed by the type of activity performed in the task. Following the familiar ideas of Anthony (*Planning and Control Systems*, Harvard University Press, 1965), these may be characterised as Operational Control, Management Control and Strategic Planning — closely corresponding to our previous classification by volume. Our third characterisation is to identify the degree of interdependency in the task being performed. Three broad classes can be identified: tasks *independent* of inter-action with any other; those requiring some *pooled* action by several people; and those occurring in a situation in which they have to be performed in a strict *sequence*. These three character-istics of office tasks are illustrated in Figure 12.3. The third characteristic of task interdependence plays an important role in the achievement of successful automation of the office.

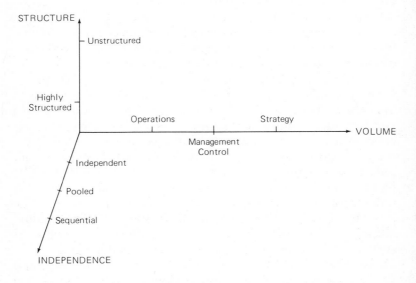

Figure 12.3 *Characteristics of Office Tasks*

Thus, the clerical functions that are directly affected by existing dp systems usually have certain qualities. For example, they are in general uni-functional; that is, the clerk performs only a single function, such as order entry. Where the clerical workers have a multi-functional role, then the different functions are usually performed in batches in sequence. This contrasts with the general nature of office work, which involves a number of independent functions, probably happening concurrently. For example, an office worker could be preparing some document and be interrupted by a request for information which involved information retrieval from a filing system. This could involve several documents being available to work on at one time, with the worker switching between them as necessary. This is very much in contrast to the normal dp approach of one thing at a time.

One attempt to measure the different tasks undertaken in the office is shown in Figure 12.4. This represents an analysis of the time spent by secretaries and managers at one of IBM's research labs. The proportion of typing is disproportionately higher than

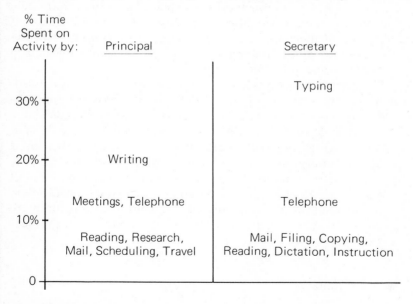

Figure 12.4 *Breakdown of Time Spent on Various Office Activities*

in a normal office due to the number of research papers. However, even in this instance it is clear that using a word-processor is unlikely to show any dramatic productivity gains. It will be necessary to provide systems which improve the productivity of a number of these activities before significant overall gains will ensue. Thus systems attempting significantly to impact global office productivity will have to support a number of different functions simultaneously, in an integrated manner.

In general, office workers themselves decide on the sequence of their work. Activities are usually event driven so that an incoming document or message initiates a procedure. There may be a number of conflicting events that cause the office worker to schedule and structure his work. This contrasts with the normal dp approach where the system is 'in charge' and dictates the structure to be followed.

Certain procedures will be explicit and laid down, but many situations will be handled informally, often by precedent — for example, 'ask Fred what he did last time it happened'. The informal networks within an office are vital for its well-being, since they represent much of the context of an organisation. It is not that such informal networks cannot be formalised, it is more the scale of them (probably exponentially proportioned to the number of people in the organisation), and their dynamic nature (they are constantly being modified).

Another factor affecting office work is that for many organisations a significant amount of the data processed comes from outside the organisation. This means that the organisation will have very little control over the format and content of the data. Human beings are particularly skilful at taking unstructured data and interpreting it. They are using not only their innate mental skills, but also their whole education, training and experience. There may be no theoretical reason why computers can't be made to perform this task equally well, but there are some good reasons why this is unlikely to happen in the near future.

Why Automate?

Having examined the complex nature of the office, one may at first wonder whether it is even worth attempting to automate this complex organism. A large body of opinion would support the view that the *status quo* has much to recommend it. Is it really sensible to spend a vast amount of time, money and energy on replacing a system that currently functions acceptably, the net effect of which will be to reduce the demand for labour even more? Leaving aside for a moment the employment question which we consider in detail in Chapter 17, there are some very strong pressures driving the movement along towards increased automation.

First, the business climate is becoming ever more hostile. The competition grows fiercer both from home-based companies, multinationals and also from overseas companies. But, being a trading nation, we have no option but to compete. Regulations grow stricter, whether it be in the areas of employment, safety guaranty or marketing. And the market place becomes increasingly sophisticated, particularly with the use of computers to collect, analyse and model data. All this puts pressure on management to increase their own productivity, to be more effective in an environment that is growing in complexity.

Productivity in the Office

One of the major factors in staying competitive is to achieve improvements in productivity — increases in manufacturing productivity over the past fifty years have been dramatic. Many organisations have reached the point of diminishing returns in attempting to improve on this. However, office costs have increased steadily over this time with very little gain in productivity. Office productivity is a difficult thing to measure, but casual empiricism would suggest that nothing dramatic has happened in the general *modus operandi* of the office since its inception, other than the introduction of the typewriter and telephone.

There is only slight evidence of a reduction in office costs arising from the introduction of computing equipment in organisations over the past twenty years or so. Usually there is a substitution of technical dp personnel for clerical personnel, with the cost of the equipment counting against any other gains. Certainly some organisations have managed to enlarge their business substantially without increasing the workforce, but usually at the cost of considerable capital investment.

And this is the nature of productivity, the substitution of capital for labour. In manufacturing this has been most effective with an average capital investment of at least £25,000 per worker. However, the average investment per office worker is probably about £2,000. The tantalising question is, can the productivity of the office be increased by investing in technology and systems? Certainly the logic makes sense — the cost of labour is increasing significantly, yet at the same time the cost of hardware is falling dramatically. The office is long on labour and short on technology; such a portfolio is seen as being less than efficient.

To summarise this argument, office costs have now risen to a point where in many organisations they often exceed half of the total overheads, and are an increasing proportion of such costs. Surveys suggest that on average roughly 10% of revenue is spent by manufacturing companies on office costs — though this might range from 3–4% for process industries, such as the oil companies, to as much as 14–15% for fabrication manufacturing companies with large sales and service operations. In the case of organisations where the office is the shop floor, such as insurance companies and many financial institutions, the average appears to be around 15% and several such organisations spend over 20% of revenue on their office operations. Further productivity increases in the manufacturing side of an organisation may be difficult to come by, so the office is an obvious candidate for attention. Given a period of slow growth, the focus is on reducing costs and the office represents a major proportion of these. Consequently this appears to be an opportune moment for such an operation.

What to Automate

The next question is, what is one trying to achieve by automation? This goes back to our previous discussion on automation *versus* mechanisation. A mechanical copy of the current system is unlikely to be effective. Similarly automating individual activities will show short-term gains, but may not help towards an overall solution.

If the objective is to reduce office costs, then it is obvious where the major part of this lies, namely the cost of managers, their salaries and the enormous infra-structure needed to support them. Other costs are usually a fraction of this, though one must qualify this when talking about a clerical office. Thus a major objective must be to increase the effectiveness of a manager.

One of the major functions of management is decision taking, and office automation must improve the quality of these decisions. However, this is more easily said than done, for the quality of decision making is a very difficult thing to measure. Indeed, the process of taking decisions is one that is very imperfectly understood, and is one on which there are several conflicting views. We have to rely heavily on the subjective view of managers as to whether this type of technology helps them to do their job better. The measurable achievement of improved decision support systems may become apparent only after some considerable time has elapsed, when it becomes possible to see that the whole organisation is functioning more efficiently as a result of office automation. Even so, productivity is a subjective measure, as it all depends on who you are and what you are measuring.

Managers require support in four main areas: first, *communications*. This breaks down into message handling, where the information needs to be delivered reliably and promptly from A to B; and interactive communications, where a dialogue is required between two, or more, parties. Messages may easily be translated into some electronic form so long as the translation process is not too cumbersome. There is much debate about whether managers are prepared to use keyboards — there is some evidence that UK managers are more reticent than their US counterparts in this. However, this can be expected to disappear when personal computers become commonplace artifacts, thus breaking the direct association of keyboard and typewriter (this can already be seen to some extent in the increasing references to 'keyboarding' rather than 'typing' skills). This may become totally irrelevant with the development of sophisticated speech recognition devices (see Chapter 14). Interactive communication, however, needs to be carried out directly and the major assistance would be in helping to make the connections and providing facilities like recorded and pre-recorded replies.

Second, a manager needs help in the *storage of information*. Information may be collected in many different forms which need to be filed away. Filing is the process of giving structure to a set of information so that subsequently the information may be retrieved relatively easily. Some of the technical problems with this are discussed in the next chapter. Another aspect of this problem is the retrieval of information that has been stored by somebody else. This could be financial data such as the IMF (International Monetary Fund) database, or any of the growing number of on-line databases available. Data storage and retrieval will have to interact with the communications facilities, so that, for example, all electronic mail would be automatically filed. Another example is maintaining a diary, where structured information about the manager's schedule is stored, which could be automatically interrogated to find possible dates for a meeting.

Third, there is a need to help with the problem of *information overload,* which is summed up by the diagram in Figure 12.5. The quality of a decision improves as the amount of information available increases from zero, and at some point reaches a maximum; but the effect of extra information beyond this point is actually detrimental. This is connected with the problem of information retrieval, but something more than a passive retrieval facility is needed. Ideally some facility should be actively scanning incoming information and deciding what is

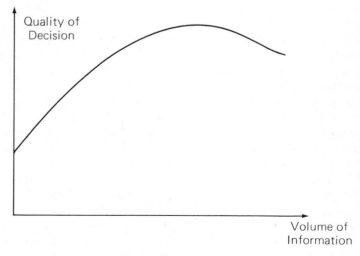

Figure 12.5 *Information Overload*

important and relevant to the manager's attention. This would require some model of the information, so that exceptions and trends could easily be spotted. Information filtering is normally done by people (often making use of machines), and after each stage the raw data used as input to that stage usually becomes inaccessible. Consequently the users of filtered data are reliant on the quality of the filtering, as there is often no way to check back directly. The filtering process is important and it is also connected with the analytical tools that a manager needs.

Consider an example of how a sales forecast may be arrived at. The salesmen give estimates of their next year's sales to their area manager. The manager revises these based on knowledge of the individual salesman, their sales performance and previous forecasting ability. The area sales forecast is then given to a regional sales manager who repeats the process. The regional forecast is then given to a national manager who again makes similar adjustments. The resulting forecast is thus the sum of a lot of different people's judgments, and any automated system would need to mirror this level of sophistication.

Finally, the manager needs a set of *analytical tools* to manipulate and investigate the information. These would enable a particular process to be modelled and various scenarios simulated to list the range of possible outcomes (sensitivity analysis or to answer a series of 'what if' questions). At a more down to earth level such tools could simply help in automating the problem of working through the numbers on a spread sheet for example, in the preparation of a budget.

Obviously a manager has to perform many other functions, but help in these four areas could provide enormous increases in management productivity. There are two essential qualifications to the way these facilities must be provided. First, all the different functions must be integrated, so that any data communicated by the system may be stored, and retrieved, and be accessible to the analytical tools. Second, the system must be easy to use. In the next chapter we discuss these two points more fully.

There is a considerable difference in the way different levels of management utilise information: senior management spend more time on planning activities, while junior management would be mainly engaged in control activities. To do this, senior management are more likely to make use of external information which would tend to be unstructured, while junior management are more likely to use structured internal information. To perform their control functions, junior management would make use of detailed internal information, whereas senior management would rely on summarised information. This is obviously an over-simplification, but it highlights the different requirements of any integrated information processing system.

Having looked at some of the requirements for the electronic office, we will now turn our attention to the way that many organisations are progressing along the path.

Towards the Electronic Office

The starting point for many organisations is word-processing, which has been 'successful' in a number of different environments. The typing pool was an obvious candidate for mechanisation and here productivity gains have often been dramatic. The size of the gain depends on the type of work — long documents needing several revisions may show gains of 5 to 1. Similar performance can be achieved where documents consist of standard paragraphs. At the other end of the spectrum, short one-off documents such as letters may show little, if any improvement in productivity.

This new technology has affected the culture of the typing pool, changing the relationship between the users, the typists and the supervisor. There are far fewer natural breaks with word-processing, so there is the danger of expecting too much continuous effort. This may be just a transitional problem and it will take some time for the culture to adapt; but a bigger question mark hangs over the whole future of typing pools. Word-processors effectively produce paper documents more efficiently. Once the facility to transmit documents electronically becomes commonplace, then the text input is liable to be done nearer the originator. There is a direct comparison with the demise of the dp punch room with the advent of on-line terminals, and one must expect a similar pattern with text preparation.

Word-processors are now becoming a standard secretarial tool. They have been particularly successful in professional offices, where standard letters and paragraphs are common and high-quality documents important. They are also used extensively in producing personalised advertising letters, though there is a growing cult of inverted snobbery which prefers hand-typed personal letters with mistakes, Tippex and all. Where secretaries are given their own word-processor, there seems to be an increase in job satisfaction. They enjoy the challenge of using a sophisticated instrument which can increase their effectiveness. However, when the word-processor is provided as a communal facility a different set of factors come into play. Secretaries don't like leaving their desks and telephones; the system may be unavailable because it is being used by someone else; and people are less happy about learning to use new complex systems in 'public' places. A consequence is that the equipment may be neglected.

Electronic Mail

All documents prepared on a word-processor can be stored for subsequent retrieval. However, all the incoming documents (mail, reports etc.) at present exist only in paper form. Ironically many of these documents will have been prepared on other word-processors where they are stored in electronic form. The missing link is some standard form of electronic communication or electronic mail. Like the traditional mail there are two different forms of electronic mail, namely, internal and external. Internal mail could be handled by the use of communication word-processors and some form of internal network (see the next chapter). The simplest way to ensure this capability is, in theory, to purchase all the equipment from a single supplier. Alternatively there are now a number of standard communication interfaces that many suppliers provide.

Internal electronic mail is relatively simple — documents need some electronic address attached to them, and the sending station needs to ensure that the receiving station is ready. Another facility that can be provided on an internal network is electronic message switching. Though no different in principle from electronic mail, it has rather different characteristics — messages are much shorter than documents and the messages often form a dialogue. Messages will usually be entered into the system by the originator directly — the manager in person rather than via a secretary. This has led to a change in style of communicating. Instead of the perfectly prepared letter exquisite in every detail, the message is shortened, misspelt and often cryptic; for example 'HOW R U', 'C U TONITE', 'PLIZ SEND ME NEW SYS DOC MAN'. This 'terminalese' is common amongst users of computer terminals, and for example, was developed to a fine art by users of ARPANET (an international research communications network).

External mail provides more of a problem because one is reliant on a third party to provide the network. In Europe this has, up to now, always meant the national PTT. In North America, however, a whole range of services are provided by different companies, from the packet-switched networks like Tymnet, to ITT's facsimile and word-processor communications networks, and IP Sharp's electronic mail system based on their time-sharing network. These networks need two major facilities: first, the ability to store the messages and documents, then forward them at the appropriate time. In this way the receiver does not need to be active at the same time as the sender. Second, it must be able to connect a wide range of different suppliers' equipment. This is one of the major stated design features of the Advanced Communications System (ACS), a new network being introduced in America by the massive AT&T (Bell) (see Chapter 10).

The term electronic mail also covers a number of services which involve the production of paper at some stage. For example there is the international facsimile service introduced by British Telecom which allows a copy of a document to be sent to several locations in North America in a matter of minutes. Another service being pioneered in the US is for companies with a large credit customer base, such as public utilities, to send out their monthly invoices encoded on a magnetic tape to the postal service. The information is sorted into regions and distributed electronically and then the invoices are printed locally by the Post Office and delivered with the normal mail.

There is already a large international network that allows messages to be sent electronically, namely the telex network, which is used widely by international organisations. It is limited in terms of speed and accuracy, but many organisations have managed to interface word-processors and electronic storage facilities to the telex network, thus using telex for external mail and having their own internal system.

This is the stage that many organisations have now reached and the question is, where to next? This must depend to a very large extent on the suppliers and the strategies they adopt, which we will discuss in the next chapter.

13

The Electronic Office

Having explored the nature of the office, we can now look at the various strategies that the suppliers are adopting. Why is it that so many different types of organisations are trying to enter the office automation market? One reason is the technological convergence that we have already discussed. Industries which traditionally were in different markets, such as computing, communications, and office products, now find that they are developing products that overlap the traditional market boundaries.

All the suppliers believe that the Electronic Office Systems (EOS) market will expand at a rapid pace. This conclusion is usually arrived at by assuming that organisations will increase their capital investment per office worker, as we discussed in the previous chapter. However, each industry has its own reasons — the dp industry is beginning to mature, with growth rates dropping to those of more traditional industries. They have a great need to find new or expanding markets in order to maintain revenue and growth, as their raw materials (i.e. the electronic components) are falling in cost every year by about 20%. Both the computer and semiconductor industries are thus hoping that EOS are going to be voracious consumers of their products.

The communications companies are in a different position, particularly with the growing wind of deregulation that is sweeping through the world. New aggressive companies are entering the market and the communications giants are finding that the previously quiet, protected havens are being disturbed by unruly competitors from other industries. The office products industry, on the other hand, is fighting for its very life and unless it makes the change to a total system supplier, it is likely to be relegated to the role of component supplier.

Widely Differing Strategies

There are amongst the potential contenders in this market a very wide range of strategies, which invariably reflect the traditional business of the company concerned. These vary according to whether they come from the computer vendors (mainframe, mini and micro), the office products suppliers, or the telecommunications equipment companies. The major variations in approach are as follows:

Distributed Data Processing	A linked network of micro/mini/mainframe computers
Local Area Network	A standard communications 'ring main' into which can be plugged work stations and many other devices
Computerised PABX	Computer controlled telephone exchange which can handle any type of different digitised signals
Viewdata	Private viewdata systems
Integrated Computer Systems	Totally automated dp systems which replace the office

DDP Approach

Many of the traditional computing companies are approaching office automation as a development of distributed data processing. There would be a network of computers set up with some kind of hierarchy; for example, a large mainframe computer might be used as a central database machine, storing all the structured corporate data. Individual applications might be processed in minicomputers dedicated to a particular use (i.e.

sales order processing, financial accounting). Clusters of intelligent terminals or personal computers might be grouped around a co-ordinating minicomputer to provide work stations for word-processing, financial planning, electronic mail, diaries, calendars, etc. This is very much a systems approach and is a natural extension of dp. The key characteristic is its hierarchical nature with clearly delineated lines of responsibility.

Local Area Network Approach

The technical characteristics of this type of network have been described in Chapter 10. The originator of this concept and its strongest advocate is the Xerox Corporation. The concept is similar to an electrical ring main that may potentially be accessed at any point in the circuit. An 'information highway' is provided around the building(s) into which can be plugged any number of different devices. The device that allows a user to perform the routine office functions is usually called a *work station*. The Xerox Star work station is one of the most sophisticated currently available. It consists of a powerful small computer, a high resolution display screen, keyboard and 'mouse' (a small device like a toy mouse that can be moved around on a desk and so direct the cursor in the screen). The user can operate the system by using the mouse to move the cursor to the appropriate symbol on the screen (filing, retrieval, messages, and so on). A number of different tasks and documents may be worked upon simultaneously.

At other points on the network there would be a file storage system to store all the documents that need to be accessed by any of the work stations. There could be a number of different printers from low speed (50 characters per second) daisy wheels to very high speed (20,000 characters per second) laser printers. There would be a communications gateway allowing the network to be connected to other networks via a public packet-switched network for example, or to connect the network to the main dp systems. Other devices could be a simple word-processing station, a personal computer, a document reader (optically reading the characters), a telex interface, and so on.

Xerox realised that a major technical barrier to the development of EOS was the specification of a communication standard that would allow many different types of devices to be inter-

226 OFFICE AUTOMATION

linked. It now appears that an international communications standard for local area networks will be formulated by the IEE (the IEE 802 standard), which will almost certainly include Ethernet as a sub-system. This will provide a common interface standard, which, if taken up, will provide the ability to interconnect equipment from different manufacturers. If the example of dp systems is anything to go by, customers initially like single supplier solutions, so that at first they will install a system containing only equipment supplied by, say, Xerox. Once they feel confident about the network's operation, then they will experiment with non-Xerox devices that will provide other or better functions. Finally maturity will be reached when a full range of plug-compatible devices become available (probably from Japanese manufacturers).

The London Business School has implemented a network based on a similar principle. However, it is rather more limited in that it allows only Apple microcomputers to be connected together. Attached to one Apple is a 33 Megabyte disk which provides a file storage capability for any of the other Apples on the network. A second Apple is used to control a printer (potentially more than one) and another Apple may be used as a communications gateway allowing the network to be connected to any other network or system (for example, to the Hewlett Packard 3000 minicomputer). All the other Apples (currently eight) are user work stations, some with their own local floppy disks, printers, voice input cards, etc. This type of system is known as a shared resource network, as it allows the file storage and printing devices to be shared between the user work stations. In addition, messages may be sent between any device on the network, thus allowing an electronic mail system to be implemented.

As we saw in Chapter 10, one criticism of the Ethernet standard is that it is a baseband system; in other words it is not designed to carry voice traffic and its band-width is too low to carry video signals. This will mean that one or maybe two physical networks will be needed if voice and video networks are to be implemented alongside an Ethernet data network. One can certainly see the economic attraction of having a single network capable of handling all communications; and Wang have proposed one solution with their Wangnet. This has a number of different communication bands along a single physical cable. One of these bands will be Ethernet-compatible, another will take video signals and it is also possible to connect telephone type devices.

At the current stage of development, baseband systems are becoming widely available, while most broadband systems are still on the drawing board. In the long run it is likely that broadband systems will become dominant; however, in the immediate future, baseband systems are likely to appear much more cost-effective. There are a number of small companies offering networks which will support a full range of communication media but whether these will survive in the long run is doubtful. What everyone wants is an agreed standard as quickly as possible, and then the problem will come down to the economics of manufacturing.

Computerised Telephone Exchange Approach

This approach is being adopted by the manufacturers of telecommunications equipment who believe the heart of the office lies in vocal communications. Some simple facilities allow more efficient use of the existing equipment; for example, abbreviated dialling codes, ring back when free, automatic redialling, call rerouting, recorded messages, etc, are all facilities available with a computerised PABX. Given the amount of time and effort involved in people contacting each other, these facilities could show some remarkable productivity gains. However, this is just the start of the kind of facilities that are possible.

In California, a company called Delphi (a subsidiary of Exxon) runs what is at first sight a glorified telephone answering service. When a caller telephones a company that is using this system, the powerful computer at the heart of the system identifies which company is being called and displays the greeting message to be given to the telephone operator. If messages may be left for more than one individual in the company, then details of the relevant company personnel are displayed. Messages can be left by either party — any message left by the caller is typed by the computer. Messages may be subsequently retrieved by the using company by calling in and giving some appropriate identification code. A caller connected to such a system would have no reason to suspect that they were not calling the company directly. This kind of service could radically change the mode of operation of a significant part of many offices.

The ultra-fast computer made by Delphi was originally licensed to Nexos for the UK market. This company was established by the National Enterprise Board to develop a British presence in office automation companies. Unfortunately NEXOS has recently been split up, with a major part of the operation being taken over by ICL, who are not interested in the Delphi licence. NEXOS were originally planning to use the Delphi computer at the heart of their system. It would do much more than provide a clever telephone service, as it could also handle all the other devices that are a part of the electronic office. NEXOS were already marketing two of these devices, a word-processor (manufactured by Logica) and facsimile equipment. Logica have reacquired the rights to the word-processor, and Muirhead those to the facsimile. The UK lead in developing an approach to office automation through a computerised PABX probably lies with Plessey, who are developing a range of office systems which will be linked through their existing PABXs. The advantage of an approach based on the telephone exchange is that it builds on the major existing means of office communications, i.e. voice, and so provides a natural evolution path. There needs to be no dramatic change but rather the gradual introduction of new facilities.

Viewdata Approach

As we saw in Chapter 10, viewdata is a simple system for storing and retrieving data over a communication network, and as such some suppliers believe it provides an excellent springboard for entering the office automation market. Private viewdata systems are now being offered by a number of suppliers as a means of making databases accessible to managers. The process of tree-structured searching leads simply (if sometimes rather slowly) to any item of data without any detailed knowledge of the data that is in the system. Viewdata is also well suited to electronic mail. However, the database updating system needs to be extended to provide a full word-processing capability, and a facility for keyword searching as in a traditional DBMS (Database Management System) is probably necessary. So far, Rediffusion has probably had most success in this area, both in selling private viewdata systems, and also in building a more general approach to office automation on this base. Currently this is in the form of

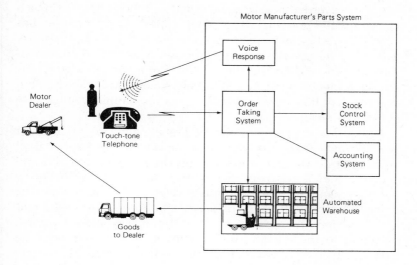

Figure 13.1 *Motor Spares System*

the 'Teleputer' — a combination of viewdata screen, mini-computer, floppy disk storage, and VCR.

Integrated Computer System Approach

This approach seeks to make the office virtually redundant, or at least the clerical part of it. It is best illustrated by two examples.

First, there is the motor vehicle spare parts system run by both UNIPART and Ford (illustrated in Figure 13.1). Each dealer who uses the system has a touch-tone phone (commonly used in the US as a simple computer terminal, each key emits a different tone which is normally used to indicate the dialling code). To place an order for spare parts, the dealer dials the appropriate computer which communicates with the user (the dealer) via a voice response system. So the computer welcomes the dealer with an appropriate greeting and asks for an identification code to be entered; this will be confirmed by the computer system by replying with the dealer's name. Stock items, numbers and quantities are then entered by the dealer, with the computer confirming vocally the type of part requested, then checking the stock file for availability. The voice response can either come from pre-recorded messages or from a synthesised voice which has far greater potential range of vocabulary. Once the order has been received it can be processed automatically. Picking lists are generated for an automatic warehouse; accounts are automatically produced and in theory the funds transfer could be automatically initiated. Not all of this system is yet working, but potentially it does have a major impact on a number of hitherto separate office functions.

The second example is in the food retailing business. Consider what normally happens when an order is placed. A supermarket stock control system would identify items to be ordered and produce the necessary paper work. This would probably be checked manually, then put into envelopes; the order would be dispatched through the print room, then to the Post Office and on to the food manufacturer. Here the mail would be opened and sent to the relevant input point to the order processing system which, if appropriate, would generate an acknowledgement. This process would continue with documents travelling backwards and forwards between the supermarket and the food manufacturer until the goods were delivered and payment made. The information flows in this process are illustrated in Figure 13.2.

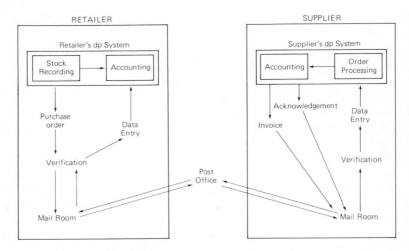

Figure 13.2a *Food Retailer/Supplier System*

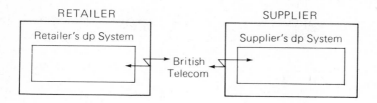

Figure 13.2b *Food Retailer/Supplier System, Electronic Ordering and Invoicing*

A trial system being introduced under the auspices of The Retail Consortium by some major supermarkets (Boots, Fine Fare, Woolworth, Tesco, Makro) and food manufacturers (Birds Eye, United Biscuit, Reckitt & Coleman, Pedigree Pet Food) is seeking to automate this process. What makes this possible is that all food items are identified by the unique Article Numbering System (these are normally displayed on the product as a bar code, a series of black and white stripes). Thus when Tesco's stock of frozen peas runs low, it will be possible for their computer to contact Bird's Eye's computer directly and initiate the order. In theory no paper work need be exchanged as the communications will be all electronic and automatic. Obviously the two parties will have laid down standards for the messages to be sent from computer to computer.

This is a good example of how traditional office clerical work may become automated. The major obstacle is agreeing the standards for the product codes and messages. The food retailing example is happening because of the high volume of paperwork between the two parties and because the common product coding, the Article Numbering System, already exists. There is no reason to believe that this will not happen on a smaller, more general scale, maybe using Prestel as a vehicle. Certainly with the advent of more sophisticated, reliable and cost-effective telecommunication networks, we can expect to see a great increase in computer-to-computer communications. This latter situation is also a very good example of office automation systems addressing a high-level strategic problem, rather than specific, personal tasks.

Technical Gaps

Even if organisations knew exactly what they wanted to do in application of EOS, a significant number of technical pieces in the jigsaw are not as yet sufficiently developed to allow for total implementation. The following paragraphs describe some of the more significant problem areas.

Work Stations

The work station is the key element in the system, since it is the user interface; that is, it is the way the system is perceived by the user — in many instances the work station is the system as far as the user is concerned. Ideally it should have full colour and graphics capabilities; it should be possible to direct it by touching (or pointing to) the screen; it needs a significant amount of data storage, which would include the capability for storing vocal messages and graphic images, and it needs the ability to work on multiple documents (split screening). While all this is technically possible, the cost of such a device today is prohibitive for general use in most organisations. The Xerox Star is a long way down this road and the terminal developed by CDC for its Plato educational system has many of these characteristics too. But the cost of these terminals will have to come down much closer to the cost of today's unintelligent terminal before every manager's desk will have one.

Document Input

One of the transitional problems of EOS is the interface to the printed word. This is a problem that will gradually diminish, but at the moment it presents a very real obstacle to getting started. There are two ways that a document may be displayed on a video screen: first as a digital image, i.e. a facsimile. The document may be displayed on the screen but the computer has no knowledge of its contents, it is solely 'a picture'. If this document is to be subsequently retrieved, then additional reference information will have to be entered with it. Second, the information on a document may be captured intelligently by an *optical character reader (OCR)*. This scans a document and interprets the information in the same way as if it has been typed in. While

OCR devices are currently well developed, there is considerable limitation on the type of document they can read; in general they can only read specially prepared documents. There is a machine developed by Kurzweil for blind people which is capable of reading virtually any book in a synthesised voice, but a book is in a very standard format compared with most business documents.

Voice Input

The most natural means for people to communicate is by talking. Already computers have the capability to recognise limited vocabularies. But this capability is very restricted — the system has to be trained by an individual for all the words (or phrases) in the vocabulary and it can only recognise words that are separated by a gap (disconnected speech). Recently IBM managed to recognise a short amount of connected speech using a limited vocabulary (less than 1000 words), but at the cost of a vast amount of computer time. Even though speech can be recognised, we are still an even longer way from understanding what has been said. For example, a computer can distinguish between 'grey tape' and 'great ape' only if it can understand the context in which the speech occurred. Another problem is the difference in sound between different people speaking; even people of the same sex, race, and locality have wide variations in pronunciation.

Much research is being done to identify those unique characteristics of words which allow us to understand one another. Undoubtedly we will develop sophisticated voice recognition systems, but these are likely to evolve gradually. A computer that can understand (not simply recognise) speech with the same facility as a human will certainly not be with us this century, and probably not until well into the next one. However the motivation to produce such systems is strong because of their universal appeal (e.g. automatic language translation), and because they will be enormous consumers of computer memory and processing capability.

Printing

Printing is probably the most advanced of the technologies involved in office automation and already provides a wide range of choice (see Chapter 11). Matrix printers are rugged and utilitarian and relatively cheap. Those employing daisy-wheel technology are high quality, but very slow speed. Laser printers are very fast and produce high quality, but currently are very expensive, though this appears to be an area in which dramatic reductions in price will be seen in the very short term. The desktop, high-quality laser page-printing system, producing 10–20 pages per minute, and costing less than £5,000, is not far away. Traditional line printers are a compromise between these two extremes. At one time it seemed that ink jet printers, which electrostatically controlled a fine spray of ink to 'paint' characters, might provide the ideal general office printer, but there have been technical complications in getting them to operate reliably. What is ideally required is a device that can print a range of character fonts and sizes as well as graphical images. It needs to handle a variety of stationery, produce good quality print and be quiet in operation. It would seem that the laser printers are the best suited, so long as reductions in price do not compromise reliability.

Data Storage

There is an enormous requirement for the storage of all types of information in machine-readable form, and quantitative data itself represents only a small part of this requirement. Textual data requires greater volumes, as does graphical and recorded voice data, and the greatest storage volumes are needed when video data needs to be stored. We have already discussed secondary storage technologies (see Chapter 3), the most important of which is magnetic disk, which still provide the most cost-effective way of storing large quantities of data. The life of these types of devices will be extended by employing optical recording techniques, rather than magnetic, which allows a far greater amount of information to be stored in a given space. Floppy or flexible disks provide an excellent storage medium in the short term.

However in the longer term, storage should be in a solid-state medium to allow for the very high volumes, rapid access time and reliability that will be required. This may be either optical, which promises very high capacity, or semiconductor which has a very fast access time. It will have to be very cheap and reliable because poor reliability causes some of today's major problems with storage devices, so that elaborate backup procedures have to be devised. If the chances of physical failure becomes small enough, then file protection can become a software problem and not an operational one. One of the great prospects for electro-magnetic storage, namely, bubble memory, has recently received a major setback as nearly all of the semiconductor companies (except two) dropped its production after considerable R&D effort. However the Japanese are still pursuing this technology, so it may still have a future beyond its present rather specialised applications.

There is another aspect of storage, which in many respects is more problematical; that is, the system by which information is indexed and retrieved. The standard technique is to define a number of *keywords* for a document which is to be indexed. For a letter this could be the date, person, organisation, and a few keywords reflecting the general content. The point about this kind of technique is that when a document is stored, judgments must be made about how it will need to be retrieved in the future.

Alternatively a *full text retrieval* system requires no prior decisions to be made when the document is stored, but searches all the text for relevant or matching keywords at the time of retrieval. This obviously places a considerably greater strain on the hardware during the retrieval process. One solution is being pursued by ICL in developing the *Content Addressable File Store* (CAFS) in which you tell the hardware what you want and it finds it for you. In typical retrieval systems the user specifies a set of keywords and the system then responds with the number of documents referenced; the search may then be narrowed down before the documents themselves are actually scanned.

Reliability

A major consideration with the design of an electronic office is its vulnerability to failure. Present offices are vulnerable to power failures, but even during these a basic level of operation can usually be maintained. Failures with an electronic office could mean the work was actually lost, destroyed or 'corrupted' (spoilt), and that the entire office comes to a standstill. If the computer breaks down, no-one may be able to process any

words, retrieve any documents, transmit any mail or whatever
— designing a lot of redundancy into the system could appear
extravagant, but in emergency it may be a vital safeguard.

For example, this was brought home to us in the summer of
1981 when the London Business School was struck by lightning.
No physical damage was done to the building, but every interface card in our Apple network was disabled, as were a large
number of terminals. One of the authors can testify to the
frustration of having a morning's work sitting in front of him in
an Apple microcomputer, but with no way of saving it.
Reliability can be designed into systems by ensuring that they
'fail-soft'. In other words if a failure occurs, nothing is lost or
corrupted, except maybe the current operation. When the
system is restarted then work can continue at the point where it
was interrupted.

14

The Challenge of System Development

Even if all these technical gaps discussed in the previous chapter were filled, it is still a debatable question as to whether we are yet capable of developing the software for a fully automated office system. For example, a recent article in *Computing* magazine suggested that at SEGAS the problem was primarily a technical one and that they had a very clear idea of what they were trying to do. Yet this contrasts strongly with many surveys both in the UK and the US indicating that very few companies have yet developed coherent strategies towards office automation. Though many organisations are in the process of doing so, several still see the best strategy to be to keep as many options open as possible.

Indeed, many organisations have difficulty in keeping up with the requirements of developing new dp systems, without any of the problems that office automation may bring. To a significant extent this has meant that the strategies are being dictated by suppliers and that potential user organisations are waiting until the shape of competitive products becomes more distinct. As we have already seen, in the Electronic Office information processing is seldom structured, repetitive, well-defined, or even remotely similar to the characteristics of a good dp application. That makes the traditional dp development process particularly unsuited; but perhaps it is worthwhile first of all to review this process.

Traditional DP Approach

The first step in the development of a system is to define the objective(s) of the system. This may not be easy if the system is needed to serve a number of different purposes, as they may be conflicting. Unless the objectives are clearly stated, there may be subsequent confusion about the performance of the system and about how to evaluate it. Next a clear statement of the problem can be made in terms of 'what it should do', rather than 'how it should be done'. Having defined the problem, then all of the possible alternative solutions should be identified and evaluated. It may be that none of the solutions is fully accept-able, in which case it may be necessary to go back and modify the statement of the problem. These stages have very often been overlooked in the past as no alternatives were allowed; for example, 'everything will be done on the main UNIVAC compu-ter', may be the only choice given!

Having decided on 'what', the focus can now move to 'how'. A system specification is drawn up which defines the inputs, outputs, database and processing steps. At this stage the emphasis traditionally has been on trying to nail the specification as firmly to the ground as possible (a proverbial analogy would be carving it in tablets of stone). Once specified, a system can be implemented in various ways, traditionally by programming in COBOL or Assembler. The system is then tested and when working to the satisfaction of the dp department, it will be passed to the user for acceptance (at which point the tablets of stone often become important!). Systems of any complexity are not liable to be error-free when delivered, and a period of live usage will be necessary to make it sufficiently robust.

Having developed a working system it must then be installed. This may involve conversions from other systems, parallel run-ning to ensure accuracy, phased implementation etc. Users must be trained and documentation provided at various levels. Some evaluation must be made as to whether the system meets its specified performance criteria and as to the success of the development process. This will be an ongoing process as the system needs to be maintained and adapted. At some stage in the life of a system it will be necessary to replace it and this will restart the system development life cycle. These stages in systems development are summarised in Figure 14.1 — or alter-natively Figure 14.2.

1. Set Objectives:
   reconcile conflicting needs;
   statement of 'what'.

2. System Selection:
   identify alternatives;
   evaluation and choice.

3. System Design:
   statement of 'how';
   data entry, data base, reports.

4. Implementation:
   programming, testing, acceptance;
   documentation and training;
   installation and parallel running.

5. Evaluation:
   systems audit and review;
   maintenance and update;
   replacement.

Figure 14.1 *System Development Cycle*

Uncritical acceptance

Wild enthusiasm

Dejected disillusionment

Total confusion

Search for the guilty

Punishment of the innocent

Promotion of non-participants

Figure 14.2 *Phases in System Development*

Limitation of Traditional Approach

There are three major principles involved in the traditional dp approach to systems development, namely:

(1) people know what they want and this can be written down;

(2) once a system is specified only minor changes are likely to occur;

(3) there is sufficient volume of data being processed to justify the cost of the software development.

Unfortunately none of these is likely to apply to the Electronic Office. It certainly does not seem practical to specify all the procedures that make up the function of an office. It is also unlikely that the problem can be divided into neat, self-contained systems with tidy interfaces to each other. The office is a dynamic environment, constantly adapting itself to new conditions and this conflicts with the static nature of dp systems and tools. The potential number of different types of transactions in an office environment is on quite a different scale to dp systems, and many of these may only occur infrequently. This might well be handled as manual exceptions in a dp system, which rather defeats the point of an Electronic Office.

The other major difference is that dp applications provide the structure within which the user can operate. There are roads and paths down which the user can travel, but they are all predetermined. In contrast EOS need to form an environment in which the users are allowed to define their own paths and in effect provide their own structure.

There is a considerable body of evidence to suggest that the traditional approach to dp development no longer functions satisfactorily. New systems are not implemented on time, do not meet user needs, or are over budget; even the simplest of modifications to existing systems appears to take months, and complex changes are not attempted; dp staff are increasingly tied up in maintenance of present systems, with fewer and fewer resources available for new system developments; the backlog of new applications already accounts for several years of development time, and is growing. Clearly, not all of these points apply to all organisations, but enough do apply for these caricatures to strike chords in many organisations. The present methods of system development have to change — the different nature of applications in office automation will hasten this process. This has led suppliers into developing methodologies and tools that are of a more general nature. Instead of trying to develop deterministic software packages, higher level application languages are needed which are non-procedural (see Chapter 4). In other words the user instructs the system on *what* is required rather than *how* to do it. As an example of the kind of systems that are beginning to emerge it is worth looking at the Visicalc system.

Visicalc

Visicalc is a program developed by a company called Personal Software, originally for the Apple microcomputer, though subsequently it has become available for most other micros. However, its initial availability on the Apple is believed to have been a major factor in their success, contributing directly to many thousands of sales. It is in effect a giant electronic spread sheet with 60 columns and 250 rows. On the visual display screen a small window is shown on this sheet — this window may be easily moved around over the whole sheet. The facilities to control this window are very sophisticated: the screen may be split into two; certain columns may be locked while others are scanned; the number of columns may be changed, and so on.

The user can label the rows and columns of the sheet with titles or other identifying information, and can enter numbers into any of the positions in it. Alternatively, algebraic relationships with other rows and columns can be specified. For example, sales may grow at 10% per year/month in either a simple or compound manner, or the unit price row multiplied by the sales volume row can produce the sales value row.

Each of these relationships is specified interactively with simple commands. The really clever part of the system is that the effect of changing any item on the sheet is immediately reflected in all the other numbers, as Visicalc remembers the formula by which each entry in the electronic spread sheet is calculated. For example, one row might contain sales volume forecasts depending on a particular growth rate; from this, profit is calculated in another row. By altering the sales growth rate, all the sales volume figures and profit figures will alter accordingly. A whole range of interactive sensitivity analyses may be undertaken. There are also facilities for calculating rates of return and present values.

After an initial investment in learning how to use Visicalc, it is quite remarkably adaptable to a whole range of problems, anything in fact that is suited to spread sheet analysis. It is used by professionals like corporate planners, by managers more used to paper and pencil or delegating the analysis, and by small businessmen like shopkeepers. Visicalc is in effect a non-procedural programming system and is a good example of the sort of tools that will become increasingly available in the future. In many ways the most extraordinary aspect of Visicalc is that it costs less than £100, as compared to a traditional financial planning package costing £5,000 or more, yet only being suited for professional use.

One of the most important parts of any system is the interface between the system and the user. This is often referred to as the man–machine interface. It is especially important where systems are to be used by non-experts as will be the case with the Electronic Office.

'User Friendly Systems'

This overworked label is much favoured by suppliers to describe their particular software. The characteristic of these systems would hardly qualify for the label of friendship in human terms, so we must view this as a relative statement which compares these offerings to their predecessors. There are, however, several observations one can make about designing good man–machine interfaces.

First, the complexity of a system depends on its functionality; the more functions that can be performed, then the more complex the interface (it is relatively simple to design 'user friendly' systems that don't do anything). Second, there is much confusion between 'ease of use' and 'ease of learning'. Certainly Electronic Office Systems need to be simple to use, but it would be naïve to believe that someone could effectively use a multi-function work station with no training. Some managers have had considerable training for the skills they need and there is no reason to believe that they would not accept this for the electronic office, if they

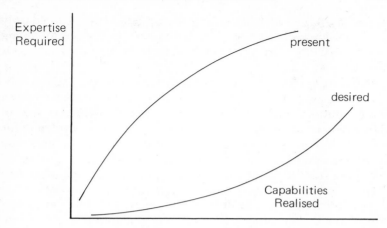

Figure 14.3 *Systems Development*

are convinced of the benefits. However the majority of managers have invested little time in training — most of their knowledge has been gained in the job, and systems for these managers will have to be easy to learn. But there is a trade-off between the level of knowledge required on behalf of the user and the degree of functionality delivered by the system. This is illustrated in Figure 14.3, and can no doubt be improved.

Next, the user should have to refer to manuals as little as possible, preferably only in rare exceptions. The system should have a hierarchy of information about its operation, and at any point the user should be able to request a list of functions that are available. There should be a consistent syntax so that the user knows the rules of any dialogue with the system and a list of the valid vocabulary should be available at each point in the dialogue.

The dialogue with the user should have at least two levels, namely novice and expert mode. In the novice mode the dialogue will be verbose, with sufficient explanation to instruct a newcomer. However, the regular user will find that this verbosity is intrusive and he needs the facility to reduce it to a minimum, possibly entering a number of items at one time, thus skipping over a series of questions. There should be a facility to put the system into reverse, to allow transactions to be backed out. A user needs to say 'oops, I didn't really mean that'. The more traditional approach is for the system to rap you sharply over the knuckles and say, 'as a punishment for not following my instructions to the letter I have just destroyed your morning's work'!

No complex system will ever be totally error-free. There will be highways and byways — the highways will be the well-trodden paths down which the user moves quickly with no fear of obstruction, while the byways will be the seldom trodden paths along which nasty surprises or obstacles may well be met. In a good system if a tree falls on a byway, it falls in front of the user, allowing steps to be retraced and another path taken, rather than falling on top of the user, causing total immobility.

A facility for users to define their own procedures enhances the productivity of a system. Users will often regularly go through a standard series of operations — if these may be specified as a single procedure, then all the user has to do in the future is to invoke the procedure rather than going through the individual operations.

All these and many more techniques besides are needed to overcome the fact that the system is incapable of conversing with us in natural language. When and if this becomes possible, systems will become truly 'user friendly'. Until that time we will have to design systems that behave in a predictable manner. This constrained and alien world will hardly be friendly, but at least it may become familiar.

Expert Systems

This search to equip our current computerised systems with a 'human window' is likely to prove totally misguided. We are currently building systems which function internally in a totally different way to the human brain. This is because of the vastly different technical characteristics of the 'pieces of hardware'.

The brain is capable of storing a vast amount of information (maybe 10,000,000,000 bits); it can retrieve virtually any chunk of this within a couple of seconds, yet its ability to 'process' information is slight. Short-term memory can hold only about 7 items, maximum processing speed is about 20 discriminations per second and internal information transfer rates can only be described as pathetic (30 bits per second). In comparison, a modern day computer can process millions of instructions in a second (even hundreds of millions). It can store millions of items in its short-term memory and transfer data internally at astonishing speeds. A decent sized dp system can even store about the same amount of information as the brain, but its ability to retrieve information is very clumsy and wooden.

Donald Michie, Professor of Machine Intelligence at Edinburgh University, believes that trying to equip our current type of system with a human face is a bit like giving the inhabitants of Dover a powerful telescope so that they can see the Eiffel Tower. Apparently sound in principle, it overlooks the small matter of the Earth's curvature. Michie and a number of academics in the US and elsewhere have been trying to build systems (and succeeding) which in some way mirror the human thought process. *Expert systems,* as they are known, are moulded on the conceptual framework of a human expert in a particular field. These systems, which are specific to a particular application, have the ability to learn from an expert, codify this knowledge, and then store it in a database. The other part of the system is known as the 'inference engine', which allows the system to make inferences from the codified data and thus hold a dialogue with a non-expert. What gives this type of system a 'human face' is that it can explain its decisions by expounding its reasoning; the user can even give the system new information to see if it affects things. This is in total contrast to a black-box system whose answers are given on a 'take it or leave it' basis.

Digital Equipment (DEC) has developed one of the first heavily used commercial expert systems. This is a system to help their engineers diagnose faults in computer hardware. The system builds on the experience gained by all their engineers, so when a new type of fault is discovered, or a new way of detecting an old problem, this information can be available to all users of the system. The system is an active, rather than a passive one — all the information contained in the database will be retrieved at the appropriate time and the engineer will not be presented with reams of irrelevant information.

Another well known system is called PROSPECTOR which is capable of helping a geologist analyse rock samples in a search for oil. This program is now being actively developed by Schlumberger in what is another example of large-scale commercial use of an expert system.

As yet these systems have not been applied to the 'office.

However it can only be a matter of time before they become more developed on well understood applications and then can be generally applied to any area. It may develop that in the future, any system that is expected to interact with a person will be an expert system or some derivative of one. This approach is also known as *knowledge engineering*.

Prototyping

One approach being adopted by several organisations in order to tackle the system development problem is the prototyping of new systems. This is being done both in an attempt to remove some of the existing problems, and to adapt to the changing nature of the systems to be developed. Prototyping is a first attempt at a system design (subsequently it may be enhanced) which is both quick and cheap.

There are several important characteristics of such an approach. First, the prototype system is an operational system — it meets some (limited) real user needs. It can therefore be evaluated against these needs. Second, it provides a mechanism to test out assumptions — both those of the user and the designer — before they are 'cast in stone'. Third, it is developed quickly — the timescale is one of days or weeks, not months or years — so that the user need hasn't significantly changed. Fourth, its development is relatively cheap — using specific development tools, rather than traditional high level languages. Fifth, the prototype system may well turn out to be the final system — alternatively, a more efficient use of resources may suggest a more professional coding. Finally, in concept the prototyping system is an iterative process — the evaluation after each step suggests a next step — abandon the system, finalise the system, or develop a further specific function.

Apart from these generic characteristics of the prototyping approach, there are three major reasons cited by users for its adoption. First, as a way of *clarifying user needs* — as we have discussed previously, the traditional system development process assumes that the user knows what he wants, and can write it down. We have already pointed out that this is a problem with dp systems, and is clearly going to be even more so in the development of electronic office systems. Thus systems development has to acknowledge that users may well be unable to specify exactly what they want. Rapid development of a prototype system may well help to clarify these needs, if only in a negative way. In some organisations this has perhaps been taken to its logical conclusion by providing tools which enable users to develop (or modify) systems for themselves. Such an approach may well, in practice, encourage users to change their minds!

The second reason is as a procedure by which the *validity of a particular system design* can be verified. For example, this can be especially useful in applications where there is a high degree of interface necessary between several systems — prototypes can be designed to evaluate alternative interface designs. A further example is in situations in which particular performance characteristics are critical, such as response times in real-time systems. Again, a prototype system can provide a means by which the performance of alternative system designs can be examined. In comparison with the point made in the previous paragraph, the points in this paragraph deal with the internal design of a system, as compared with the external design described previously.

However, the third reason cited by users is that with which we began this discussion of prototype systems — namely, an effective process for the *successful creation of a final system*. It is this iterative process which lies at the heart of the prototyping approach to system development. Essentially it can be envisaged as a way of breaking down a potentially large development problem into a number of smaller entities, each of which fulfils some need in its own right, and can be individually evaluated. It presupposes that the interface standards can be specified. Although the development of the final system may appear suboptimal as compared with a single, global development process, this presupposes that a global approach is feasible at the outset, not just with hindsight. However, the overriding advantage of the prototyping approach is its flexibility to modify, or even abandon the overall development strategy. Furthermore, the user receives some supporting system which he can use on a much earlier time-scale. This gives him confidence about the success of the overall project, and tends to make him its advocate, rather than detractor.

One thing to remember is that prototyping requires different software tools to enable the designers to create systems rapidly and cheaply. Most of these we have reviewed already: database management systems, application development systems, expert systems, and so on. Not only does it require different software tools, it also requires different skills on the part of the system designers and users. The designer requires the skills of a business, rather than systems analyst. The development procedures too are different from those described earlier in this chapter. As explained earlier, the significant characteristic is the iterative nature of prototyping.

Pragmatic User Strategies

Whilst there are many theories and concepts as to the appropriate next steps in the development of office automation in organisations, these are often perceived by practising managers as being an ideal, but unattainable situation. So it would appear useful to review what pragmatic steps may be taken by an organisation wishing to take the next steps. Five major steps may be identified, based on concepts typically applied to strategic choices facing an organisation.

Critical Success Factors

Critical success factors are a familiar concept to those involved in formulating strategic business policy — they are those few key areas of the organisation which ultimately *must* be got right if it is to be successful. For most organisations critical success factors are few (typically in single figures), and are already identified and generally accepted within the organisation. It is important to appreciate that these factors are in the context of the purposes of the organisation as a whole, not of the dp or information systems department. This step should identify the areas in which efforts should be concentrated, ensuring that the goals of office automation are supporting the overall goals of the organisation.

Needs and Requirements

Having identified the critical success factors for the organisation (or the part of it under consideration), the next step is to examine how improved information support systems could meet specific needs and requirements. Generally, this is accomplished most easily by consideration of the specific needs of the managers concerned. The particular needs of managers that might be addressed have already been discussed in detail in Chapter 12 — namely, communications, storage and retrieval of information, information overload, and analytic tools. The balance between these needs and the specific systems will obviously be determined by the characteristics of the organisation. However, what we must not lose sight of at this stage is the overall critical factor being addressed — specific tools for managers are a means, not an end. Formal procedures have been implemented in some organisations to help identify these needs. Typically these are based around the concept of *Information Resource Management,* and involve current and prospective information flow and requirement surveys, in the form of questionnaires, activity diaries, personal interviews, statistical analysis of file and task activity rates, and so on. Often a useful by-product of such activity is some more coherent view of the current status of automation in the organisation's procedures.

Future Goals

Trying to predict the future, especially in a situation where technology is changing rapidly, is clearly risky. Too often the process is focused closely on the changes in the technology, rather than on the overall system that may be required. Projections are probably best made in the form of scenarios describing how a particular portfolio of technologies might be applied to meet a specific need. These scenarios would include a consideration of the context and environment in which the desired system was to function, both in terms of organisation structure as well as human factors. The aim would be to examine the characteristics of the technologies required for the system to be a success, perhaps the two most important characteristics being price and performance.

These scenarios also have to consider how the functioning of the organisation may change if the systems envisaged in the scenario are indeed successful, and also if they are not. Some examples of such scenarios can readily be seen; for example, the clearing banks may well have as a future goal a system in which personal customers can initiate (and complete) a whole range of transactions from their own homes, perhaps using some view-data system. Customers could examine balances, initiate and modify standing orders, order foreign currency, order cheque books and (eventually) make payments. Such a system is techni-cally feasible now, and may well be cost-effective in the not too distant future. But its implementation would clearly have a major impact on the bank's organisational structure, yet meet the needs for reducing operating costs, providing a better service to customers, and improving competitiveness.

Thus, so far in this procedure, critical factors in the success of the organisations — and specifically of a target group of managers and/or processes — have been identified; for these factors specific needs and requirements have been itemised; and, finally, some alternative future goals have been examined, typically in the form of future scenarios. So, how do we progress from the current situation to meet these specified needs? How do we achieve the chosen scenarios?

Building on What You've Got

Most organisations have already taken some tentative steps towards office automation, usually by mechanising particular office tasks, especially word-processing. As we have discussed earlier, however, the entire development of commercial data processing since the mid-1950s has involved the automation of more and more office functions. Thus most organisations already have several data and word-processing systems, which typically operate independently of each other. Indeed, many organisations have difficulty in imposing common standards on their existing dp activities. For these, the next immediate step is normally to bridge the gap between these existing systems by linking word-processors to form electronic mail systems; developing new networks using facsimile technology and so on.

Building on what is already in place can save in terms of design and development effort, and usually appears as the lowest risk alternative since the steps involved are typically small ones. It may be possible to implement new systems without making major changes to existing sytems, and building on their strengths. However, there is also the danger in this approach that it may confine developments in areas which are inappropriate. It is important that the previous step of examining future goals should remove any blinkers imposed by existing systems. Many of the products, especially software, presently coming onto the office automation market are particularly targetted at filling in some of these gaps in existing systems.

As we have described above, a natural strategy for the suppliers of electronic office systems is to build upon their existing market strengths. Thus the computer companies are selling ddp, enabling their customers to build networks linking their present dp systems. Now many of the computer companies are also trying to sell their customers word-processors and electronic mail systems to operate over the same network. Similarly, the telecommunications companies build voice and text messaging systems on their PABXs, and the office products suppliers offer local networks of their word-processors, reprographic systems, file stores and so on.

Developing Alternative Plans

In our earlier discussion of system development, and in particular our discussion of some of the problems that users encounter in the use of computer-based systems we employed the analogy of a highway system: if you stray on to unused by-ways, you may encounter some surprises. In our earlier analogy, we suggested that in complex systems users would inevitably venture down such by-ways sooner or later, and that a characteristic of a good system was that it enables the user to extricate himself safely from such a situation.

As the final stage in our development of a practical plan for the successful implementation of electronic office systems, we would suggest that the user organisation attempt to chart out both the highways and the by-ways facing the organisation in its attempts to adopt such systems. The danger is that the new systems will be seen only in terms of the broad highways opened up by them. The implementing organisation should endeavour to make sure that there are no nasty surprises lurking down the by-ways.

The challenge is to create plans which map out these alternative routes and their consequences, and minimise the nasty surprises. There are several rules of thumb which might be followed: progress should be in the form of short steps, building on established systems. Each new step should be seen as justifiable in itself in the facilities it provides. Unless vitally necessary in terms of the critical factors, the organisation should not be involved in leading edge technology — established, well-understood technologies should be used. It will be important to establish organisation-wide standards where appropriate — especially in communications and user interfaces. Systems should be developed only with full management support in the user departments, and so on — the usual criteria for successful systems development apply. However, perhaps more than present computer-based systems, the new electronic office systems will have greater potential impact on organisational structure and the role of the people who make up the organisation. Thus creation of user standards, support, documentation, education and training is as important as the design details for each system, but an overall concept of the goals of the business and the place of electronic office systems in supporting their achievement is critical. This is the objective of the steps set out in the previous paragraphs.

Conclusions on OA

In conclusion, one might consider the comments of P.A. Strassman who once worked in the Xerox Palo Alto Research Center (PARC). He observed that the office of the future is a label which hides more than it reveals. It is certainly not explicable solely as a technical phenomenon but is better understood as the beginning of structural changes in the job relationships amongst 'knowledge workers'. We can observe some very strong pressures working to make it a reality. Organisations certainly have an increasing problem in the operation of their offices. They would like to be able to substitute capital for labour, thus reducing their overhead costs and increasing their effectiveness. However, in general, they have no clear idea how they are going to do this, but rather are relying on suppliers to provide systems that have the desired effect.

The suppliers believe that if they can just get the bandwagon rolling, it could mushroom into the world's biggest industry. Unfortunately there are still a number of technical gaps which they are doing their best to paper over by designing systems which fit the available technology rather than suiting the user.

Even when technology closes these gaps it is still not at all certain that we have the skills to build the software systems necessary.

It is an area that nobody can afford to ignore but at the same time everyone is very wary about making the wrong move. The danger of technological *cul-de-sacs* is very real and the memories of similar disasters with dp systems are fresh in the minds of both users and suppliers. One thing seems certain: one day in the not too distant future IT is going to happen, and when it does those left out will really be in the cold.

Part VI

THE IMPACT ON SOCIETY

The preceding chapters of the book have mainly concentrated on the technological, business and marketing aspects of IT. What differentiates IT from previous major advances in technology (e.g. electricity, railways, telephone, television and nuclear power) is that while the technology is still in its infancy, we can foresee some of the effects it may have on our society. We do have options for the future, but these need to be brought to the wider attention of the public. All too often this is done by the media in such a way as to induce fear (through ignorance), or amazement at technology's magical powers. We need to realise that we have options for the future, and to appreciate that now is the time to consider them, before too many doors have been closed.

In Part VI we will be looking at three major issues and their impact on individuals, organisations, and society as a whole. First, we look at the related and interlinked questions of privacy, data protection and security. Next we consider possible changes in the nature of work, and in particular its implications for training and education. Third, we look at the potential impact of IT on aggregate employment and the creation of wealth and we constrast the approaches taken by different national governments in their policies towards IT.

15

Privacy, Data Protection and Security

Privacy

Privacy has always been a central political issue in any society. It is possibly one of the major distinctions between the Western democracies and the Communist bloc. It reflects the relative importance of the individual within society, as compared to the state. In Western society one's privacy as a person is considered a basic human right and typically is extended to encompass your own private space: 'an Englishman's home is his castle'. However, the privacy of one's living space will depend greatly on the wealth of the individual. Living in a small flat in a high-rise block scarcely affords the same privacy as living on a 5,000 acre estate.

Another aspect of privacy is the status of personal information. Who does such information belong to — to the individual (to whom it relates), or the state (as custodian of society's rights), or the public at large (through the media)? In this aspect the influence of wealth may have the reverse effect. Information regarding the personal conduct of the wealthy and famous is of far greater interest to the newspapers than that about the poorer members of society.

In the UK the control that the individual has over the dissemination of this information is mainly through the laws of trespass — the situation may differ in other countries. This potentially allows false information to be corrected (for those who can afford it), but in no way does it restrict the flow of correct but personal information, or even enable the subject to know of its existence. However, as these laws were not designed with the protection of privacy as their objective, it is hardly surprising that they don't meet the need. Societies have always kept records about their individual members for administrative purposes. The abuse of these files has generally been kept under control by two factors: a professional discipline imposed locally by those responsible for collecting the information and, possibly more important, by the inefficiencies of the filing system used. With the advent of the electronic computer the second of these controls no longer holds true.

Computerised Name-linked Files

It is interesting to consider on how many computerised files, information about you, personally, may be stored. For example, your employer's personnel and payroll files, public utilities (electricity, gas, water), local authorities (rates or local taxes), maybe government files such as social security, inland revenue and the health services (soon in the UK). UK car owners are recorded at the DVLC at Swansea, which in turn makes the information available to the Police National Computer (PNC) at Hendon. If you have ever reported a crime to the British police, then details will probably be held on the PNC — at present the PNC holds files on 22 million people, more than half the adult population. Your bank will have you on file, as well as your credit card company. So will one or more of the credit checking agencies. The credit card company may have passed your name on to a direct mail agency, or the direct mail agency itself may have picked up your name from the Electoral Role or one of many other sources of names and addresses.

None of this is necessarily particularly sinister in itself. If the computerised databases were self-contained boxes that could be isolated and only accessed locally, then we could possibly rely on local controls to regulate them. However, with the development in telecommunications which make it simple to access a computer remotely or to link computers together, we can no longer totally rely on these local controls. There are now beginning to emerge many examples of situations in which these databases have been used in a manner for which they were not intended. One example was in the Ladbrokes Casino licence court case during which it was revealed that the car numbers of customers of the rival Playboy Casino were noted by a police officer. Using his authorised access to the Police National Computer, the police officer obtained the names and addresses of these people, who were then duly invited to the Ladbroke Casino!

Perhaps an even more disturbing example emerges from the West German police's attempt to fight the Bader Meinhoff terrorist movement. In Germany, nearly all libraries have a computerised lending service. Thus it was possible to discover the names and addresses of all people who had taken out certain kinds of book (such as those by Marx, Trotsky, etc.). Combining this list of people with other information, maybe from school records, housing records, etc, it was possible to identify a section of the population who were likely suspects. While many people would have sympathy with the end, are we really prepared to countenance the means? And if we are in this specific context, how can we ensure that in another context, of which we did not approve, we could exercise any control or influence over such activity? Such powers of search and surveillance developed to fight terrorism could be employed by governments of a different complexion to control and manipulate the population.

What makes this a particularly important and relevant issue now is that we can actually do something about it. The remote and linked use of computerised databanks is in its infancy. But gradually working practices are springing up and once these become established it will be very much harder to change them. In comparison with the current public interest in the nuclear debate, there is very little political interest in the UK about this subject. Will we be seeing mass demonstrations against computerised databases in twenty years' time when it is too late to do anything about them? If we care about the kind of problems we are storing up for future generations we should consider the implications of turning every telephone exchange into a computer, when experts confidently expect computers to be able to recognise (and understand) general speech by the end of the decade. It will be easier if any necessary legislation is in place before the technology develops any further, and is formulated from principle rather than expediency.

Data Protection

The privacy of personalised information held in computerised databases is often referred to by the term Data Protection, and in many Western countries the rules and safeguards of Data Protection have been established through legislation. Probably the most advanced country in this respect is Sweden. Here anyone who wishes to set up a name-linked computerised file must first register it with the Data Protection Agency. This is an independent body set up by the government to monitor and enforce the workings of this legislation. So far over 100,000 files have been registered in Sweden which are subject to the provisions of this legislation.

The legislation covers three major areas. First, is the data on the computer *accurate?* This is answered by giving each individual the right to know exactly what information is stored about them. The Data Protection Agency has the powers to enforce database operators not only to divulge the contents of their files, but also to correct mistakes and inaccuracies. Certain files are too sensitive for the eyes of the general public, but if an individual suspects that incorrect information is stored on, say, the police system, then a member of the DPA has the right to inspect those files and act for the individual if they are not in order. In the UK, although you can find out most of the information about you held on a computerised database with varying degrees of difficulty, correction of inaccurate entries is very difficult, relying on expensive litigation.

The second aspect is that of *authorised usage.* Information collected for one purpose should not be used for another, without the express permission of the individual(s) concerned. Thus for example, it would be illegal for a credit card company to pass on your name to a third party for marketing reasons without your explicit permission.

Third, the *age of the information* in the database may render it irrelevant and/or inappropriate. Information will stay stored in computers until it is physically deleted. Information about the events of yesteryear must either be deleted after the appropriate time, or only be displayed with the date of the event being clearly shown. For example, certain court convictions should be deleted from the record after a certain number of years has elapsed, but this is not an automatic procedure and has not always been carried out.

The Swedes are a regulated society and this kind of legislation comes more easily to them. But it is not just the Swedes who are taking action. The Council of Europe, of which the UK is a member, have produced the Strasbourg Convention on Data Privacy, which all member states have agreed to ratify. In ratifying this convention the UK has committed itself to introducing some kind of legislation. Unfortunately the story so far in the UK in this area is rather dismal. In 1972 the Younger Committee on Privacy highlighted all these problems and suggested that legislation was necessary. This was quietly forgotten about until 1978, when the Lindop Committee on Data Protection came out with firm recommendations for a *Data Protection Agency* (DPA). Since this time there has been a certain amount of muttering in corners, but it appears that action in this area has amongst the lowest priority of any prospective legislation.

The British apathy to this problem (there are no votes to be won on this ticket, which accounts for the politicians' lack of interest) presumably stems from a great faith in the fair play of the British system. This faith is beginning to be broken down by an increasing number of revelations of misuse. The dilemma facing the current UK government is, first, an aversion to regulatory legislation (their whole philosophy is one of deregulation); and second, even if the regulations are laid down, how are they to be administered? The government has been busily demolishing as many 'quangos' (Quasi Government Organisations) as possible, so it has no great desire to establish another bureaucratic agency like a DPA.

The government's current proposals are for some data protection agency, which would be administered by a 'data ombudsman' and a small staff. However, the codes of practice they would administer would be voluntary, and not a mandatory part of the legislation as in other European countries. Furthermore, the ombudsman could act only upon receipt of a complaint from the public and would have no power to initiate an independent investigation.

These are the proposals at the time of writing; but, by the time that legislation finally reaches the statute book, they may well have changed. However, as we shall discuss in the next section, other pressures are forcing the government to speed up their legislation timetable in this area. All of the current proposals from the government exclude the government's own database operations from the remit of the proposed agency, particularly those administered by the Home Office. Indeed, in some of the proposals, the data protection agency would for administration purposes be a part of the Home Office — this would almost certainly mean the emasculation of such an agency, given the record of the Home Office in suppressing the rights of the individual in recent years, and its almost pathological insistance on veiling everything in a web of secrecy. It would appear to us that only a data protection agency enforcing mandatory codes of practice on all users of name-linked files can adequately guarantee the rights of the individual.

However, what is most likely to pressure the politicians into action is the problem of trans-border data flow.

Trans-Border Data Flow

This impressive sounding phrase refers to the passing of personal computerised information over national borders in electronic form. Certain countries (such as Sweden and Holland) are bringing in legislation which will prohibit organisations transmitting such data to countries which do not have adequate data protection legislation in place, such as the UK. This will be very inconvenient for some multinational companies, such as Shell and Philips for example, who maintain company personnel records in both the UK and the Netherlands. So we can expect that all the major Western trading nations will enact suitable legislation. This is the major source of the pressure on the UK government to enact DPA legislation.

Presumably this will give rise to *data havens*, where more unscrupulous operators will be allowed to store whatever data they like. Thus a credit reference agency could operate an unregulated off-shore service, though there is some debate about how they could manage to do business. Another ironic side effect is that the Third World countries have woken up to the potential of this legislation as a means of controlling the multinationals. So again it is likely to be the multinationals who will be the front runners in pushing for change.

Data Protection versus Freedom of Information

On the face of it there would seem a potential conflict between these two campaigns. The freedom of information movement is about greater access to government-held data, but in doing so there must be some protection for the rights of the individual. Thus in many countries these two campaigns are linked together — although curiously, in the UK this strong link does not exist. The UK has a long tradition of closed government, and the forces opposed to this may spring from a more radical stream than the liberal tradition of protecting the individual. Ironically, in the UK, those arguing in favour of more open government have sometimes seen Data Protection legislation as an obstacle to their aims (since it would give the individual 'obstructive' powers). However, so far the protection of the integrity of those operating the British public services has been put forward as a major objection to freedom of information.

Security

The final aspect of this issue concerns the security of information from the viewpoint of an organisation. In many organisations information is rapidly becoming one of its major resources. It represents the knowledge and nervous system of the organisation. For example, recently British Petroleum observed that they could afford to have an oil rig, production well, refinery or tanker out of service for some time; but the one thing they could not afford to be without was their computing systems. Computer systems are becoming a very sensitive and vulnerable part of an organisation, as the UK government recently found out at great cost. A strike by a few computer operators almost totally paralysed its revenue-collecting mechanisms. In Italy the urban guerillas have discovered that blowing up an organisation's computer centre is the most effective method of causing damage (far better than kidnapping the chief executive!).

Computer Fraud

Being highly dependent on information in order to control an organisation also makes this process vulnerable to incorrect and misleading information. Given that an organisation takes many actions based on information received only in electronic form (the banks electronically transfer $100 million daily), then protecting systems against fraud becomes vital. The scale of computer-based frauds is a much debated point — many organisations don't like to think about it too carefully for fear of what they might find. Certainly the scale of crimes discovered so far is fairly impressive.

The 100 or so cases reported last year involved a total fraud of over £50 million. Many experts consider this to be just the tip of the iceberg and estimate the total level of fraud could be twenty to thirty times this amount, since only the really big frauds seem to come to court. Probably the most famous computer fraud was the Equity Funding Scandal. This company, in an effort to increase its growth rate, created phoney insurance policies that were sold to re-insurers. It was only practical to use a computer to perpetuate the fraud on the scale they did. In total some £70 million of phoney policies were created and at the trial about one-fifth of this was attributed to the computer. In this case the computer helped to increase the scale of fraud.

Recently Wells Fargo charged Ben Lewis, an operations officer, with a fraud valued at over £10 millions. This had been done by a series of fund transfers using the computerised inter-branch accounts settlement system. But possibly the most cavalier case is that of Stanley Rifkin, who defrauded Security Pacific of £5 million in a single transaction. Previously a computer consultant for the company, he obtained the funds transfer code and then initiated a transaction from a public call box, transferring the money to a waiting Swiss bank account. Here it was exchanged for diamonds which Rifkin promptly brought back to the US. He was caught only by his indiscreet behaviour.

However these are only the 'glamour' crimes. They are uncovered because of their scale rather than any particular control mechanism. In thousands of smaller companies petty (and not so petty) fraud is taking place. One expert observed that most of the people who had been caught had either behaved foolishly or had not used very clever systems — the mind boggles at the thought of what the clever criminals are up to.

There are two other types of fraud an organisation may experience, apart from illegal funds transfer. First, there is electronic espionage in which confidential company information is obtained electronically. In the US an oil company that was narrowly outbid on many oil lease sales subsequently discovered that its data line from its Texan Computer Centre to its Alaskan terminal was tapped. Many companies now keep all their quantitative information on a computer — business plans, market forecasts, product prices, technical specifications, market research and so on, representing a major asset of the company that can be devalued by disclosure.

Another type of fraud involves unauthorised usage of some electronic service. The most frequent targets of frauds of this type are the time-sharing computer bureaus who are engaged in a constant battle against the illegal user. But with the growth in high value-added computer-based services, such as investment modelling and on-line databases, this is becoming a more serious problem.

The White Collared Criminal

A fairly clear picture emerges of the computer criminal. For example, Donn Parker of Stanford has studied many cases, from which he concludes that the offender will be young, highly motivated, bright and usually an employee (past or present) of the defrauded organisation. Many organisations are unwilling to prosecute the guilty parties if they are discovered, partly to avoid adverse publicity and partly because court proceedings are such a minefield where computers are concerned. Even when criminals are convicted, the penalties are modest compared to those handed out to the more traditional criminals. In the UK the Great Train Robbers were given sentences of up to 35 years for stealing two million pounds. Admittedly the train driver was violently attacked, but it is doubtful if the sentence would have differed greatly if there had been no violence. Compare this with Stanley Rifkin's sentence of three years.

In Western societies 'white collared' crime is still considered to be somewhat socially acceptable. Those involved often do it to 'beat the system', and consequently trying to improve the system to make it foolproof is only likely to increase this motivation — the challenge is simply that much greater. There is also a belief that defrauding large organisations is acceptable — a sort of Robin Hood philosophy. Only a change in the law classing computer fraudsters with traditional criminals can lead to a change in social attitude to white collar crime. It may be that the computer criminal will always have a Robin Hood image in the public eye.

There are also a number of factors that are tending to increase the problem. First, the availability of personal computers: in the US there are over half a million in operation, and this figure is expected to grow tenfold in the next five years. With one of these machines it becomes practical to break coding schemes on electronic messages or break password systems. Second, the number of people who understand computing is increasing, fuelled as much as anything by the same growing popularity of personal computers. Third, computer systems are becoming more accessible, with organisations installing terminals, remote access facilities and electronic offices.

All this is compounded by a certain ignorance and naïvity amongst senior management. The company's computer systems are very often viewed only in technical terms, and the technical staff are forced to make many decisions which should have a wider participation. The pressure on developing new systems means that anti-fraud measures may be viewed with hostility, as they can slow down the development process. In order to build secure computer systems the programmer has to think like an auditor — unfortunately one can't think of two greater extremes in mental approach. Systems are obviously most vulnerable to the programmers who create them, and one essential control is a separation of responsibility between development and operations. It would appear that in the long run a realisation by senior management of the critical business role of corporate data and information processing systems is the only way in which fraud will be recognised as a serious threat.

16

Work, Skills, Education and Training

The Nature of Work

The whole history of civilised man has been a constant struggle to improve productivity, to move away from subsistence living where every second of every day had to be purposefully employed simple to stay alive. Gradually we have had to work less hard for the things we need and want. But suddenly in this last part of the twentieth century we find that we can now harness technology to such an extent that we apparently have labour surplus to our requirements. This leap in productivity is not likely to be short-lived, as that most productive of all technologies, the microchip, has only just started to be deployed on a wide scale. In the next chapter we will discuss the wider issues of aggregate employment; but first we will look at the nature of work and speculate on what changes IT may bring to it.

More than anything, work today symbolises a place to go which is separate from the home. This separation of work and home occurred with the industrial revolution. The new manufacturing industries required large labour forces that worked together in specialised factory buildings. These industries soon built up large offices where white collar workers were required to assemble every day. With limited communication facilities, it was considered only practical to run an office under a single roof.

Separating the work and the home has served to emphasise that work occupies the major part of our waking lives. We probably spend about the same amount of time either working (and travelling), or awake and not working. Of our non-work time maybe half is spent doing necessary things in life — washing, dressing, cooking, eating, shopping etc. This leaves us with maybe one quarter of our time to spend at our own discretion. It is not surprising then that work is the largest single thing we relate to. We relate to people through our jobs; when asked, 'what do you do?' we reply that we are engineeers, mechanics, postmen or undertakers. This describes the major part of our lives and the fact that we like gardening or squash or Beethoven is of secondary interest. An alarmed reaction is hardly surprising then if we are suddenly confronted with the possibility of totally altering the balance of our lives. If we only worked for half as long as our disposable leisure time then the question 'what do you do?' has a different answer.

Whether we need to work to have meaningful lives is a difficult question to answer. But it would seem logical to argue that members of a society will feel more valued if they are needed by society, and that in turn will lead to more social behaviour. However, the question becomes more focused when we consider whether society has the obligation to provide that work in the form of employment. Nobody is disputing the right of anybody to work, what is being disputed is the obligation of the state to provide jobs for everyone. The responsibility for work is thus transferred from the state to the individual — a concept well recognised in the US and currently being emphasised by the present UK government.

Employment is in practice discouraged through a series of legislative and fiscal measures. There are labour taxes, employers' insurance contributions, health and safety regulations, pension rights, employment protection, and a host of other measures. All of these are excellent and commendable in their own right but they exert a negative influence on the willingness of organisations to employ people. The net effect of this may be to change organisations from operating in an environment in which they directly employ people to perform tasks, to one in which they enter into contractual arrangements for the provision of services. Thus many people would become self-

employed, and could potentially work for a number of organisations. This does not mean everyone working in cottage industries, but it might well require the establishment of community
work centres where individuals could share the cost of facilities
such as telex, electronic mail, photocopying, computer work-
stations etc.

The Wired Society

Developments in the equipment that may now be attached to the
telephone network mean that not all office functions need to be
retained under one roof. Thus the office may be distributed into
smaller local units without any reduction in the communication
capability between these units. These local units may actually be
people's homes. It is fairly common for regional sales representatives to be based at their own home, with only occasional visits to
a regional office. This is now becoming practical for a wider
range of occupations. With the generally increasing cost of transport, the economics of using a terminal and a telephone link are
becoming more and more favourable. It may mean that extended
sessions using a terminal are best left to the off-peak charging
hours, but this emphasises that working at home means
scheduling one's own time. With the widespread availability of
microcomputers then all the necessary information can be con-
veyed on floppy disks; for example, 10 large (8 inch) floppy disks
could hold the equivalent information to that contained in the
paper files in one filing cabinet.

For example, with an Apple computer costing less than £2,000,
one could have the facility to use Visicalc to analyse a variety of
different types of data, and the capability to store any number of
models and data on disk. These models could even be submitted
to one's boss in floppy disk form! A whole range of word-
processing software would be available which could be used to
revise a report which had already been entered in draft form.
The Apple could access any of the time-sharing bureaus via a
telephone line to extract data from an on-line database, or alternatively to access the company's main dp files. It could also
access Prestel, extract relevant pages of data, and store these on
disk. These and a host of other facilities could mean that many
'knowledge workers' would actually be better off working from
home.

However, the concept of home-based society, communicating primarily in electronic form would seem a little unrealistic. In his book *The Wired Society*, James Martin considers the telecommunications networks to be tomorrow's highways. Instead of people travelling to communicate for whatever reason (work, shopping, services), all their communications could be electronic, and any physical goods could be brought to people rather than *vice versa*. Whereas Martin was only exploring this idea, it does need tempering with a little realism. Not everybody finds their home a desirable place to spend a majority of their time — many people find the physical working environment a preferable place to spend their days. On top of this the pressure on many family relationships would be intolerable — most relationships need the external influences of separate experiences that going to work offers.

What would seem more likely then is that an increasing number of people will work partially from home. These will be knowledge workers of some sort with a high standard of education and motivation. But for the rest of the work force, having some separate and gregarious work place will continue to be essential.

The Balance of Skills

Organisations are finding that IT changes the range and scope of the skills they need in their workforce. In pre-computer days an organisation would have a continuum of clerical skills, from apprentices, juniors and trainees to senior clerks and supervisors. Computer applications often occupy the middle ground of the skills range. This means that workers may either be button pushers, with a totally mechanical knowledge of the system; or they may have a conceptual understanding of the system and what it does, which enables them to use the system rather than be used by it.

The skills required for these two roles are very different. At the mechanical level, speed, accuracy, stamina, and the ability to follow rules, are all desirable, while creative and inquisitive skills may be a positive hazard. In contrast, at the conceptual level a worker must have some comprehension of the capability of a computer, the ability to conceptualise, make inferences, be creative and enquire. These kinds of skills are usually acquired at school and through higher education, not through company training of any sort.

Skills at Risk

What happens to workers in this middle skill range where IT is applied? This is perhaps best examined by use of a few examples. Many organisations who supply equipment containing complex electronics have been concerned about the escalating cost of maintenance. In the days when the electronics consisted of discrete components, then faulty equipment either had to be returned to a repair centre, or else a highly trained engineer had to try to diagnose the fault and repair it on site. These maintenance engineers needed to have a high level of training in electronics and have an intimate knowledge of the particular piece of equipment in which the fault lay. Nowadays most of this complex electronics has been reduced to a small number of circuit boards, if not a few chips.

Now no attempt is made to repair boards on-site; instead boards are replaced until the fault is corrected and then the faulty board is taken back to the repair centre where automatic test equipment can diagnose the problem. In an increasing number of situations it is becoming possible for the equipment user to do the maintenance by keeping spare boards, or by using in-built diagnostics which pinpoint the error, thus allowing the faulty board to be exchanged, say, via a courier service. None of this augurs well for the future of electronics maintenance staff. In one sense we can view this as a deskilling process, but in another we can look at this as being the natural consequence of change. No longer are the skills associated with understanding discrete components necessary; however, the demand for those who can work with boards as components is increasing sharply. The big question is, what happens to those who are threatened by this change? Does the desirability of their skill levels decrease, or do they acquire new skills to replace their obsolete ones?

To take another example, the group of workers possibly most at risk to deskilling are middle grade clerical workers. Such workers would have a good 'mechanical' knowledge of that part of the clerical system with which they interact. Older workers may have worked in several different departments, and thus have acquired a wide range of experience. What value will these skills (and this experience) be if the clerical functions of the organisation are automated — and if so, will the loss of the workers' experience be a significant deficiency in the automated system?

The functions in a new automated system are unlikely to resemble their manual predecessors. Thus these middle grade skills are likely to be no longer required. Low grade skills will be necessary to feed the data into the system and generally to follow the procedures laid down by the system. High grade skills will be necessary to comprehend the concept of the system and its functions. These skills will be needed for supervision, control, maintenance and development. But the middle grade skills are liable to become redundant, with the only option open to workers, other than redundancy, being to work at the less skilled levels.

One of the great disillusions of IT is that there is a glamour and sophistication associated with it. Thus workers are often led to believe that their jobs will be more demanding and interesting. While this is the great promise that this technology holds out to us, too often the reverse is true. Take word-processing, for example: at first sight it can appear to a typist quite exciting. It requires learning new skills and it may be environmentally a more pleasant piece of equipment than a typewriter. However, this rosy view may be somewhat dimmed by the reality of operations. The word-processor has been acquired to increase productivity, i.e. to do more typing per day. Though it may be easier to correct mistakes, to go on to the next page etc, these activities used to provide natural breaks. Thus the typists can find themselves in a production line regime where the only breaks are the scheduled ones.

Skills in Demand

What skills will we need for the future? In the short term we will need many 'production line' clerical workers (though not as many as at present) to operate the new electronic offices. These people will have keyboard skills and be able to follow rules and procedures (often quite complex ones), though the demand for these skills will diminish in the medium term as office automation proceeds. Much of the data will be entered automatically; for example by electronic ordering, financial transactions, funds transfer, etc. The major skills that are likely to be in short supply now are those of the knowledge worker; someone technically literate with strong conceptual skills. Such a person has only to know that the possibility of some action exists, to find out how to do it. Though it is not necessary to be highly numerate, a logical mind is essential, so a course in philosophy could be an excellent training!

The demand for skills in designing and developing IT will increase for the foreseeable future. In applying the technology in products, it is necessary to be familiar with both the hardware and the software so that the appropriate balance can be struck. Our current training schemes tend to separate these two areas. In contrast, applying IT to a process places a strong demand on systems thinking. At present hardware and software are often totally separate developments. The computer industry is often talking about chronic shortages of staff; but no longer are there major shortages of programmers (except in specialist areas), though the market rate would drop a little if supply was increased. But there is a major shortage of systems designers, systems analysts, and business analysts — the people who translate the business problem into a computer solution.

A great weakness in the UK labour force has been a lack of well-trained managers, and particularly ones who are technically literate. Technology has always had an 'oily rag' image in the UK which has discouraged generations of potential managers from learning about it. By the time that they realise they need this knowledge it is often too late, and all the technical decisions have to be left to experts. Management may be totally at the mercy of technical experts who effectively make the decisions. This contrasts strongly with American, Japanese and European managers who have a strong technical grasp of the processes and products relevant to their organisations. The training for many American managers is a first degree in engineering followed by an MBA, and it is noteworthy that the largest class of first degrees of the entrants into the London MBA program is in engineering.

Though there will be an increasing demand for knowledge workers and managers, the large reduction in clerical workers will leave a significant part of the labour force to be redeployed. Unless they are soaked up in government bureaucracies these people will have to learn new skills associated with one aspect or other of the service industry.

Education and Training

As we shall argue in Chapter 17, the UK lacks any clear policy towards IT, and nowhere is this more obvious than in its policy towards education and training. Instead of a coherent plan based on some projection of the level of skills required, there is a series of piecemeal projects each designed as fire-fighting excercises to quieten down criticism.

In secondary schools it took two years for the Department of Education and Science to agree what form their microcomputer project should take. Even then they had not considered how the schools were going to acquire the necessary hardware (this was forced on the Department of Industry). Where are the forecasts for the number of students taking computer science 'O' and 'A' levels over the next ten years? There are still more students taking 'O' level geography than computer science. Does the government think that this is a sensible state of affairs? If not, what is it trying to do to change the situation?

The strongest part of the UK educational system in respect of IT is probably the polytechnics. In the universities considerable sums are spent on research into computing and on providing computing as a service, yet the number of computer science degree places is still surprisingly low. Is this because organisations consider the knowledge gained through such a course not particularly desirable? Or are such courses not providing the right kind of training? Even more extraordinary is that the recent cutbacks in higher education spending have actually reduced the number of places in electronics and computer science, while at the same time the government exalts the virtues of IT and provides other funds to promote education in this area. As we shall argue in the next chapter, an educated labour force exploits technology, an uneducated labour force is exploited by it. In this context it is amazing that the UK is now the only industrialised country in which higher education is not available as of right to all those qualified to receive it.

We have already discussed education as an application for IT in Chapter 8. It is unclear what skills we are trying to teach children. We have our own views, which we expound in the following paragraphs, but what is of greater concern is that there is little sense of urgency in the government or the various educational establishments to formulate or discuss policy in this area. Microcomputers should become an everyday part of school equipment and all children should be capable of using them. It may be more effective for a school to buy, say, 100 Sinclair ZX81s than spend the same money on two sophisticated micros. The second of these options will undoubtedly encourage the bright and motivated children, but may well exclude the majority through the simple fact of limited availability. Having a large number of simple machines, however, would encourage (enforce?) all children to use them at the expense of frustrating the brighter children. The talents of the gifted can be encouraged through computer clubs or access to local technical colleges.

Using computers can encourage children in the pursuit of numeracy. Maths can seem a very dull and irrelevant subject taught by blackboard and textbook. Computers bring numbers to life. For example colour graphs, histograms, pie charts, simulations or animations can illustrate mathematical relationships and operations. Teaching teenagers to use Visicalc would provide them with a useful skill and a practical insight into mathematical model building. In higher education we need to concentrate more on system skills, problem analysis and logical reasoning. In addition more attention needs to be paid to communication skills such as graphic arts, data presentation, report writing, interviewing, negotiating, etc.

But separate from the skills required to apply IT, there is a far bigger re-orientation needed in our view of education itself, and our perspective on its role in society. In the past we have believed in the concept of a 'job for life'. This has allowed us to concentrate our educational process in one massive shot during childhood and adolescence in the hope that this will be sufficient to set us going on some life-long course. It is now being realised that part of the inflexibility of most European work forces is due to this concept being reinforced by our schooling, both by the content and by the length of time invested.

If we are now moving to an age when we can expect to have a number of different jobs or careers, then we have to take a different attitude to education. It must become a process that continues throughout our lives, both formally and informally. Instead of being educated until the age of 16, 18, 21, 23 or whatever, we should consider giving young people work experience before they continue to higher education. University professors have often observed that degree courses were more effective in the days of national service, when students had a slightly wider perspective than that of the average school leaver.

Continuing education and vocational training will have to become the norm if we are going to cope with the fast changing world that IT brings. In the US there is a much greater tradition in this field, partly due to placing the responsibility for employment (and thus skills) with the individual. In Japan this process is handled by the larger companies — in guaranteeing 'employment for life', the company can re-train its labour force as necessary without generating fear or resistance.

17

Employment, Economic Activity and Government Policy

An Historical Perspective

Economic Activity

Many commentators have suggested that the impact of semi-conductors, chips, and the other building blocks of information technology will lead to such fundamental and far-reaching changes in the fabric of our society that it may be thought of as being akin to a second industrial revolution. The first industrial revolution applied machines to amplify the power of men's muscles, the second will apply IT to amplify the power of men's minds. In this chapter we examine some of the consequences of this possible second industrial revolution, particularly in its impact on employment and overall economic activity.

In the first industrial revolution, the machinery of the great Victorian innovators and entrepreneurs was used to increase the power and effectiveness of man's muscular and manipulative skills by several orders of magnitude. In the first half of the 19th century, total industrial production doubled roughly every twenty years. During this period it was what we would now regard as the basic heavy industries (e.g. coal and iron production) that were providing this expansion, the major consumer products being largely those of the textile industry. In general, a worker's total income went in satisfying his and his family's needs for food, clothing and shelter.

However, the end of the century saw the beginnings of the consumer-goods industries which have dominated the economic development of the 20th century — radio, telephone, domestic electrical equipment, and, above all, the motor car. The historical development of the automobile industry is the only one even to closely match the explosive growth rates of the semiconductor industry. For the years from the turn of the century through to the end of the First World War, the American automobile industry roughly doubled its production every two years. Most other consumer industries increased their output 8 or 10-fold in the first few years of their existence, but then settled down to a more leisurely rate of growth. It is perhaps worth repeating that the present semiconductor industry has doubled its output rather more than every other year for the past 20 years, and looks set to continue at this rate for most of the next decade. Such a continuous high rate of growth is unprecedented and unique.

Employment and Productivity

Using the new machines not only increased production dramatically, it also made the population much better off. Referring to Figure 17.1 it can be seen that over the years 1851 to 1911 real income per head in the UK rose by roughly 80% and over the years 1911 to 1961 by roughly 100%. At the same time there were consequential changes in the pattern of employment — between 1851 and 1911, employment in agriculture fell by 22%, whilst in manufacturing it rose by 99%, at a time when total employment itself rose by 96%. Thus the increased use of machinery resulted in a rapid increase in manufactured output which not only absorbed all of the considerable increase in the working population, but also led to a significant exodus of workers out of agriculture into manufacturing. Not only was there an increase in output per head in manufacturing industry, there was also an enormous increase in agricultural productivity — fewer and fewer agricultural workers were feeding more and more manufacturing workers.

Looking now at the period from 1911 to 1961, there has been a much smaller growth in total working population (31%), but much bigger shifts in employment patterns. During this period, the number of employees in agriculture fell by a further 41% and the numbers in manufacturing fell by 0.7% while those in ser-

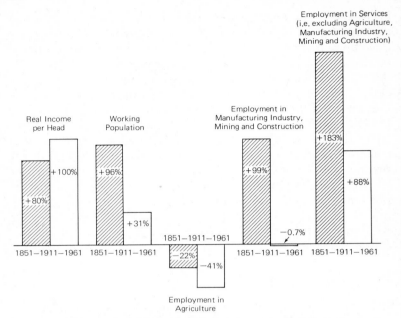

Figure 17.1 *UK Changes in Employment and Output*

vices (including finance and commerce) rose by 88%. By 1961, only 4% of the UK labour force was employed in agriculture, compared with 49% in industrial and 47% in non-industrial employment (excluding agriculture). Again, there were dramatic increases in productivity during this period — fewer and fewer agricultural workers were feeding more and more workers in the other sectors, and a relatively small increase in the numbers of industrial workers were producing the goods demanded by the vastly increased number of workers providing services to the community. Over the years a smaller and smaller proportion of the average worker's total income has been required to provide the basic needs of food, clothing and shelter, and an ever increasing proportion is available for discretionary spending on other goods and services. Such a shift in employment patterns — reduced employment in agriculture, a small increase in industrial employment, and a much larger increase in service employment — was common to all industrialised countries during this period.

The Impact of Employment

During the 1970s all of the industrial countries have experienced some significant degree of structural unemployment, seen as a shift out of manufacturing employment. Whilst some of this shift has been the result of developments in technology, almost none is directly attributable to the impact of microelectronics.

The immense productivity gains that are likely to be achieved by the adoption of microelectronics and its application, whether of the factory or the office, and the impact of information technology in general, must lead to changes in the structure of industry and in aggregate employment patterns. Much of the public comment about the impact of microelectronics has been negatively concerned with the possibility of increasing unemployment (understandably), rather than with a positive view of the new opportunities opened up by this technology. A common fallacy in such comments has been that there is only a finite amount of work to be done in the national economy, and that if machines do more of this fixed amount of work, then obviously there must be less left for humans to perform. From this come the ideas of ever increasing leisure (and unemployment), and (hopes?) of eventually working only a one or two-day week.

However, hopefully a moment's thought will make it clear that man's wants in terms of either goods or services are by no means satiated, or even nearly so. Certainly they may be more closely satiated in the case of specific goods in particular markets; for example, most people in the industrial countries do receive sufficient of the basic foodstuffs. But even when considering fairly commonplace commodities, most households would desire more of most items if the price were right. For example, although most households do possess a television set, the advances in technology that have led to the production of portable sets at lower prices have opened up new markets for second and even third sets in many homes. Thus, as we have attempted to point out earlier in this book, particularly in Chapter 8, the adoption of microelectronics in existing products can lead to dramatic improvements in price/performance, which may considerably expand the market for those products, or lead to innovation into completely new products. Five years ago the market for domestic video cassette recorders didn't exist.

Given the present labour-intensive nature of offices, any improvements in productivity in the office environment might lead to considerable unemployment. The big question is, will this unemployment be temporary or permanent? Where will these people, in addition to those unemployed because of the effects of productivity improvements in manufacturing industry, find new employment? Some commentators have suggested that in 30 years' time the total UK production of goods will be manufactured by only 10% of the labour force. Often this suggestion has been made as an indication of how terrible the impact of microelectronics will be (with the implication that the other 90% of the labour force will be unemployed). Yet it should be regarded as a great potential benefit that machines might provide so many of our needs.

Examination of previous changes in employment patterns suggests that rarely, if ever, has mass long-term unemployment occurred as a direct consequence of technical changes. Indeed, rather the opposite would seem to be the case. Relative technical competence would appear to be strongly associated with real differences in living standards between countries. The evidence available would suggest that the employment effects are relatively small and of short duration when domestic firms take the lead in technical innovation. In this situation overall growth in the economy is stimulated through the multiplier effects of the increased output. However, the employment effects are likely to be substantial and long-standing when foreign competitors adopt new technologies faster than domestic producers and, through lower relative costs and/or improved products, dominate both domestic and third-party markets. The major national concern in the UK should be with competitive price pressure from technologically superior foreign competitors. This potential threat again emphasises the critical role of the educational level of a country in furthering technical under-standing and creating positive attitudes to technical change.

Patterns of Employment

To a great extent these possible developments in employment patterns are a continuation of the long-running trend in employ-ment patterns already mentioned. The long-run decline in agricultural employment, and increase in output and in productivity has now been going on since the middle of the 19th century, yet the world's farmers are feeding more people at a higher standard. As the productivity of the agriculture worker has increased, so has his level of skills, to the extent that now the average American farmer possesses a first degree. Similarly, there is evidence to support the contention that the countries who have adopted new technology fastest are those whose standard of education of their labour force is highest. To repeat, an educated labour force exploits technology, an un-educated labour force is exploited by it. Given the experience in agriculture over the past century, suggestions that the proportion of the labour force employed in manufacturing may fall to 10% in 30 years from now, and yet produce a vastly higher output, do not appear to be unreasonable, and should be welcomed.

All the signs are that more and more people will be employed in the service sector (including commerce and financial services, and central and local government), which raises two particular points for discussion in the UK context. The first is the recent suggestion that the UK has too great a proportion of its labour force engaged in essentially non-productive industry (particularly government); and that (i) this is higher than that of our competitors, and (ii) it would be of benefit were it to be reduced. Figure 17.2 shows the proportion of the labour force employed in the agricultural, industrial, and non-industrial sectors (excluding agriculture) for both the UK and our major competitors; it doesn't appear to be immediately obvious from this that the proportion employed in the non-industrial sector in the UK is significantly out of line. Further, it is also not immediately obvious that reducing this proportion would be of benefit.

In recent months there has been much comment to the effect that things would somehow be better in the UK if more people were employed in manufacturing — and indeed, all recent governments have spent immense amounts of public money to try to maintain employment in manufacturing industry. Switching workers from government employment to manufacturing may simply worsen the productivity of both sectors. There is no evidence that manufacturing industry is short of labour, yet there is evidence of unsatisfied demand for the services that government provides.

For example, the actions of the last Labour government in removing the direct grant status schools from the state educa-

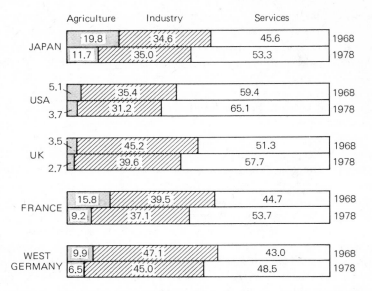

Figure 17.2 *Percentage Employees by Category, 1968 and 1978*

tional system, and the subsequent success of these schools in the private education sector, demonstrates a demand for perceived high-quality education. Similarly, the closure of private beds in NHS hospitals, rather than sounding the death knell of private medicine, has proved to be salvation, identifying a considerable market for such services. There is also plenty of evidence to indicate that such demand is justified, in that our standards of health and social care have fallen significantly behind those of the other major industrialised countries. Similarly, a more highly educated population would better be able to cope with and turn to its own advantage the immense potential of the availability of cheap machine intelligence. A relevant economic question, beyond the scope of this book, is whether a mixed private/public educational, health care, or social service system is the best way to provide such services to the community in general.

This brings us on to the second specific point concerning the switch of employment in the UK to the service sector; that is, the disastrous productivity record of the UK economy, and our apparent inability to innovate to the point of successful market exploitation. As has been discussed previously, adding value in the semiconductor revolution is found to be highest at the point of application to the final consumer — both private and corporate — and during the 1980s an entirely new integrated information-processing industry will develop, potentially dominated by the present computing and electronics companies.

At one point the UK government showed itself to be aware of this trend by the creation of INMOS (semiconductor manufacturing) and NEXOS (office systems) as subsidiaries of the NEB. While INMOS was set up with the tangible and immediate objective of producing chips, it was never clear what role NEXOS was designed to play. NEXOS has now been dismembered, and INMOS, after a period of vacillating support now appears to have established itself, although its longer term future may not lie in government or UK ownership. In our view one of the more telling criticisms of recent UK governments is that they have no coherent policy addressing the impact and role of information technology in general, and its role in the innovatory process in particular. Thus, we will devote the remainder of the chapter to examining the process of innovation, particularly in information technology, and compare the situation with regard to government's policy in the UK towards information technology with that in other industrialised countries, especially Japan.

Aggregate Economic Impact

Balance of Trade

If we examine the recent performance of some of the major sectors of the UK electronics industry in a worldwide context, we find it fairly uninspiring. Some statistics are shown in Figure 17.3. In order to give some indication of size and importance, the percentage of world imports and exports provided by the UK in 1970 and 1979 are shown for each sector, together with the size of the world market for the products of that sector in 1979.

Thus, although we have just about managed to maintain our share of computer exports, our share of imports has risen dramatically, largely because we failed to develop a successful domestic minicomputer manufacturer to compete with the likes of DEC, HP and DG.

| Product Classification (SITC code) | UK Share (%) | | | | World Market Value | |
| | Imports | | Exports | | | |
	1970	1979	1970	1979	1970	1979
Office Machines (714)	12.6	12.9	8.6	10.7	7214	19911
Computers (714.3)	4.4	14.2	14.9	13.7	988	10031
Telecommunications (724)	4.3	5.5	7.1	5.3	4179	24321
TV Receivers (724.1)	3.6	5.4	2.0	2.8	715	4606
Radio Receivers (724.2)	3.0	8.7	0.5	1.0	929	5694
Domestic Electrical Equipment (725)	4.8	6.8	8.0	4.5	1327	7176
Electronic Measuring and Control Equipment (729.5)	8.3	10.8	11.1	10.3	1040	6444
Transistors etc. (729.3)	9.0	5.8	5.8	4.3	1378	12278

Figure 17.3 *UK Share of World Trade in Electronic and Related Equipment, and the Value of the World Market (US $millions)*

In the case of telecommunications equipment, our import share has remained roughly constant, but we have steadily lost our export markets. The reason in this case is probably the non-availability of electronic, digital exchanges from UK manufacturers, until the very recent announcement of System X. At present the really big market for telecommunications equipment is in the less developed rather than industrial countries of the world. This is because the industrial countries already have a large, complex telecommunications network which they can update only slowly over a fairly long time-scale. Several less developed countries, however, are going from a basic system in one or two major cities to a nationwide satellite or microwave based digital system almost overnight. In the instrumentation industry the big development recently has been the explosive growth in 'smart' instruments. As the data in Figure 17.3 illustrate, we have also lost out in this market, maintaining our export share (largely by the activities of the UK subsidiaries of the major American manufacturers), but increasing our share of world imports.

Our export share of transistors and other components (this category includes integrated circuits and the other basic electronic components) has roughly been the same, whilst our import share has declined significantly. There are two ways of interpreting these figures. One is that our share of imports of the basic raw material for electronic components (and the vast majority of chips are imported) has fallen relative to our competitors. They are using more and more electronic components in the products that they produce. An alternative explanation is that these figures merely reflect the activities of the UK subsidiaries of the American semiconductor manufacturers. They have been able to satisfy UK domestic demand, thus reducing the need for imports. However, this latter interpretation is perhaps naïve, as until recently output of chips in the UK was low and a large proportion is exported (to some extent because of a lack of domestic demand). Most UK chip manufacturers suggest that chip utilisation in the UK lags significantly behind that of our major competitors.

UK Economic Performance

The impression that this review gives is that the UK economy appears to be incapable of adopting new technology as rapidly as our major trading competitors, whether the technology applies to the production process, or to the final product itself. Obviously there are notable specific exceptions that can be cited to this broad generalisation, but at the aggregate macro level it holds true. Thus we lose out in two ways. The machines and the process we use in the production process are not as technically advanced as those of our competitors, with the result that we employ far more men to produce the same goods; and second, the goods that we do produce are technically inferior to those of our competitors. And in many markets, particularly the high value-added final consumer markets, we simply don't have a significant domestic UK producer on a world scale.

The market for these technically advanced products is therefore satisfied by increasing imports, whilst our exports suffer.
Our total share of world trade in manufactured goods has been
in decline for several decades now. Thus UK manufacturing
productivity suffers by not producing high value-added goods,
and using inefficient methods of production for the goods that it
does produce. The problem is compounded by the predilection
of many UK manufacturers for badge-engineering. Not only
does this practice provide the foreign manufacturer with access
to a UK distribution network and marketing name, which would
have been extremely difficult for the foreign manufacturer to
achieve otherwise; but it also can mean that the UK manufacturer
makes it much more difficult for itself ever to get back into local
production of the product concerned. This is because each
pound of sales revenue from the UK provides a percentage
(typically around 10%) to support the R&D activity of the foreign
manufacturer; as the foreign manufacturer is likely to be already
technically ahead of his UK counterpart, this means that the UK
manufacturer is helping to perpetuate this situation.

It is true that the above analysis is perhaps exaggerated and
somewhat simplistic, but it conveys the significant features of
the macroeconomy over recent years. Examples of this can be
readily found: how many of the electronic toys and games sold
this Christmas were manufactured in this country? How many of
the video cassette recorders (excluding UK packaging and
badge-engineering)? The two deficiencies of the UK response to
technological advance compound each other — the labour shed
by the introduction of more efficient productive processes
cannot be employed in new high value-added industries where
the value of its product is larger, as these are the selfsame
industries that we have failed to develop. And in this time of
economic recession, it is the high technology end of the
consumer market which is continuing to expand. As a trading
nation, we have to acquire the capabilities to produce some high
technology products and services and to automate some
production processes, otherwise we will have fewer and fewer
products and services to trade in exchange for those which we
don't choose to produce for ourselves.

Opinions on the impact of information technology in general,
and the cheap, pervasive, intelligence of microprocessors in
particular, range from the wildly optimistic to the equally wildly

pessimistic. The optimists suggest that we are about to enter a new age of prosperity in which robots will produce an infinite range of wonderful new goods; products and services will be personalised; and we will all have ample time to lie on the beach. The pessimists agree that the robots will indeed produce the goods, but the problem will be how to give the unemployed the money with which to buy the goods produced. Both views are extreme, but both contain a part of the truth. Which of these extremes we actually end up closest to is not dependent upon the gods as some appear to believe, but is to a significant degree under our collective control.

Impact on Manufacturing Industry

Before exploring some of the policy options available to the government to meet some of these problems, it is perhaps worth examining some overall characteristics of a manufacturing organisation in order fully to understand the place and role of microelectronics and information technology. The product of a manufacturing firm, as delivered to the final customer, often contains a relatively small value-added share attributable to the actual manufacturing process itself (i.e. casting, forming, shaping, cutting, joining, etc.). A rapidly growing share is attributable to packaging the product and getting into the final consumer's hands (i.e. distribution, advertising, etc). A large and also growing share stems from conceptualising, designing and differentiating the product relative to its competitors. Further, for a growing number of products, information, support, maintenance and insurance (guarantees) are a significant part of the product definition. Thus, for example, a survey of American durable-goods manufacturers suggested that less than 50% of their employees were engaged in direct production activities. As a result many large organisations are as much marketing and distribution organisations as production operations, and the impact of information technology is as much on these non-production activities, as on the production processes.

However, this does not mean that the product itself, or the processes employed in its creation can be ignored or relegated as being of peripheral significance. All of the other functions of the organisation derive their strength and purpose from the competitive characteristics of the product and the processes used in its manufacture. What this does imply is that an organisation's investment in buildings and machinery is no longer solely the critical part of its investment strategy. The large and increasing service and support component is also critical — investment in marketing, design, R&D and human resources. Increasingly, the technological component in the design, production, marketing, distribution and support of a product or service is the critical competitive weapon in the hands of many organisations. Many organisations, whose primary business is not of itself technically based, are discovering nevertheless that the application of appropriate technology is a significant competitive factor in delivery of their product; for example, the travel industry, banks, and insurance companies, are all discovering this. And for some industries, most notably the entertainment industry, information technology is blurring the distinction between products and services; for example, films may be shown in cinemas, broadcast on public or pay-TV, sold or rented on video cassette or disks.

It is important to appreciate that the two types of technical change (i.e. in the products themselves and in the processes used in their creation) are often mutually dependent. Commercial success stems from and is dependent on changing products and product design. New products typically lead to the introduction of new production processes — the two are seen as part and parcel of the same decision. As an explanation of technical change this is probably much more important than a simple desire to uprate production techniques. Surveys have illustrated that many of the individual machines used in the production process are very old; for example, surveys in the mid-1970s showed that in the UK and USA roughly 60% of the machine tools in use were over 10 years old (compared with 30% in Japan). Examination of the process of technical change at the level of the firm shows that in recent years attention has been concentrated more upon how the flows of production are organised, rather than on the capabilities of the individual machines used. Studies have shown that in an engineering shop most productivity improvements have come from developments in production scheduling, rather than from improvements in individual machine performance.

Advances in technology enable us to produce more goods (both currently existing and new) and provide more (existing and new) services with the same input levels — thus productivity rises. Higher productivity results in lower unit cost, producing either higher profits, lower relative prices, or both. Consumer and corporate real income rises, leading to an increase in overall demand, whilst labour demand falls and, in perfect markets, so do real wages to maintain equilibrium. Insomuch as the new technology allows new products and services to be provided, this generates investment, and increased output and employment in the economy that produces them. Similarly, some existing products, processes and services will be rendered obsolete, with corresponding reductions in output and employment in the economy producing them.

What is clearly critical is, first, the speed with which the multiplier effects on aggregate output work through the economy, so that the labour displaced by higher productivity (assuming unemployment rises, and real wages don't fall) is re-employed in those industries benefiting from the increase in aggregate demand; and second, the mix of industries present in our economy, between those created by the new technology and those rendered obsolete by it. If our mix tends towards the latter type, then increased demand will simply create wealth in other economies whose mix tends towards the former type. Thus in order to minimise temporary unemployment we require that the benefits of productivity gains be passed on as rapidly as possible, and preferably in the form of lower relative prices. To minimise long-term structural unemployment, and to have any chance of improving our real wealth, we must develop industries exploiting new technology products, processes and services.

Innovation and Growth

The capabilities of microelectronics and information technology enable superior production machines to be developed, but, much more important, enable complex production processes to be modelled and controlled. In principle this has been possible for many years, but recent developments now make it possible in practice. This focuses our interest on the innovation process as being *the* critical factor in technical and ultimately commercial success. Innovation is the process of bringing developments in science and technology out of the laboratory and into the market-place, as *successful* products and services, or productive processes. The emphasis on successful products in this definition is of paramount importance — the process of invention creates new products (and processes), innovation further develops them into successful products. This stresses the importance of innovation not only in the creative stage of a new product, but also in the making and marketing stages. The UK has an enviable reputation in invention — the creation of the advances in science and technology — but an even more unenviable one in innovation.

In one sense the problems of the UK economy described above might be seen as a part of a national economic lifecycle — under-development, followed by a period of very rapid growth, then a slow down into maturity, and finally stagnation in old age. Historical examination of ancient civilisations — Chinese, Persian, Egyptian, Greek, Roman — bears this out to some degree or other. It could be argued that the UK was an under-developed country up until the end of the 17th century, when industrial development began, followed by the first industrial revolution in the 18th and 19th centuries. In this period the UK economy went through an extremely rapid period of industrial development. But by the second half of the 19th century this had slowed down, and in the present century the UK economy has been in its mature phase. One could suggest that the UK economy is now into its old age, incapable of movement or change, and will be moribund for a century or so, by which time it will be an underdeveloped economy again, and start the cycle over once more. Some evidence for this is shown in Figure 17.4 — since the end of the first industrial revolution in the UK, other industrialised countries have been slowly overhauling us.

However, we would regard this as being a very pessimistic viewpoint, and would hope that there is more than one way for an economy to cope with old age — hence, the importance in our eyes of focusing on innovation as the process by which our economy remains vital and capable of adapting to a changing world, even in its old age. And looking at those areas in which innovation is critical, we hope that this book has illustrated that the development and application of microelectronics seems certain to occupy a central role in the future development of the economy.

The emphasis in this area is less on creating new scientific or technological advance, but more on effectively using those that

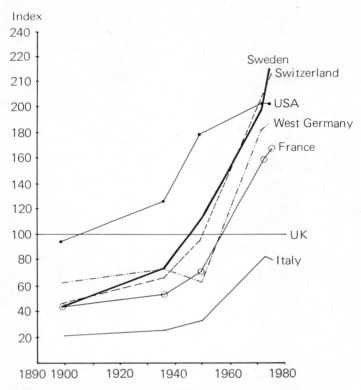

Figure 17.4 *GNP per capita in different countries, 1890—1975*
Index UK = 100

already exist. The key in this process is that well-known economic phenomenon, the entrepreneur. This appears to be one area in which his or her talents are critical — the role of transferring technology, modifying or creating organisations, developing markets is, at its most successful, a very personal process. Why is it that this process appears to be unsuccessful in the UK? Amongst the more significant reasons have been suggested the following:

> the apparent aversion of UK management to technological risk taking, and its preoccupation with short-term firefighting, rather than longer term strategy;

> its lack of real technical expertise at the highest levels, and the lack of technical competence or awareness in general management;

the outdated craft union structure, dividing the organisational objectives of management and labour;

the lack of sufficient and/or appropriate financial resources
and incentives;

the long-standing tendency of the UK educational system
to favour pure science at the expense of applied science and
engineering;

and a lack of coherent and consistent government policy to
support innovation in general, and for microelectronics
development in particular.

Perhaps at a generic level the fundamental reason is an
aversion to change, by both the organisation and the individual.
This can manifest itself not so much as an unwillingness to adopt
new products and processes, but more as an unwillingness to
give up the current products and processes — an unwillingness
to unlearn. Old ways are best, it has always been done this way,
so why change now? Products only change when market forces
make it inevitable, by which time it's too late, employing
obsolete technology and out-moded machines because they still
have a few more years' physical life left in them. There is a lot of
evidence available to suggest that the UK lives more in the past
than other countries, reliving past glories, and unwilling to
accept new realities.

No one country or company has a monopoly in resistance to
change, and, as we have mentioned previously, a good example
is provided by the major valve manufacturers and their attitude
to the new semiconductors. Their response was to try and
produce a better and cheaper valve, not because they couldn't
understand the new technology, but because they couldn't
change rapidly enough. The established valve manufacturers
had too much to lose and forget — the newcomers had very little
to lose.

Research and Development

The engine that drives innovation is R&D, which has three major
purposes for an organisation, namely:

to develop existing products and services — if possible to
lead the competition;

to improve the range and scope of the organisation's products and services to cover new market opportunities in its main business area; and

to identify and develop new products and services with market potential (possibly unrelated to existing activities), to provide a portfolio of new business opportunities.

This suggests that innovation through successful R&D implies market leadership. Yet many UK companies argue that to follow others into new markets is the best strategy. However, we would suggest that a strategy of following, rather than leading, has several potentially disastrous consequences, especially in such a rapidly changing area as microelectronics. In particular:

the first company to introduce a new product or service may benefit from novelty (gimmick) value;

the second (or later) entrant is likely to have to face the mark 2 version of the leader's product;

the leader chooses the time and place for the action, while the others may have it chosen for them;

the followers may have to follow a market development strategy dictated by the leader, which is unlikely to be to their best advantage!

A naïve model of technical change, the mechanism of innovation, suggests the following steps:

basic research creates ideas for technical progress;

of these ideas, a proportion encourage applied research, which in turn encourage development projects;

of these projects, a proportion are taken up by the firm into production as new products or processes, which enables new products to be marketed, or existing products produced at lower cost, or some combination of the two.

This simple one-way model would suggest that more expenditure on R&D is a 'good thing', since it would create more ideas for technical progress. But such a model is both oversimplified and distorted; one of the key issues in R&D is to choose the right problems to work on, and, in an industrial context, this requires a better understanding of actual and potential problems in production, marketing, distribution and all of the other areas relevant in delivering a successful product or service to the final customer. This implies a strong feedback loop in our model of technical change, and the integration of the R&D activity with the mainstream production and marketing activities of the organisation. This also has implications as to the nature and number of the skilled human resources needed in this process, and would suggest that in the UK context too many scientists and engineers are engaged at the creation of ideas stage, and too few at the make and market, implementation, evaluation and feedback stages.

The Process of Innovation

This discussion of the mechanism of innovation is in general terms. We would suggest that its importance in the area of information technology is critical, given the pervasive nature of IT and its potential influence on products and processes in all organisations. Loss of leadership in IT would appear likely to accelerate the decline of the UK economy into senility — a positive, determined approach to retain a significant design and production capability in information technology products would appear likely to be the best opportunity of giving the UK economy an active and vigorous old age.

The ability to absorb new technology and exploit its potential is based on several important factors, one of the most significant of which is the level of education of the work force. The level of literacy, numeracy and technical understanding will be a major determinant of their capability to understand change, and people's willingness to accept and actively foster change is strongly influenced by their capability to understand it. If people cannot cope with numbers, follow sequences of instructions, and infer logical conclusions, or interpret and understand diagrams, then technical horizons are limited. But education in these skills, above a certain level, enhances people's ability to absorb and adapt, to change and grow rapidly. Observation suggests that the proportion of people reaching particular levels of academic achievement in the UK are substantially less than in other industrialised countries.

There are many situations available to illustrate that once a technological breakthrough has been made, a newcomer to the technology can reach an equivalent level of competence faster than the originator. (It takes less time to get to the corresponding level, but the newcomer is unlikely ever to close the gap with the originator.) The educational level of the population has a major influence on the time it takes the newcomer to become competent in the new technology.

Thus, the ability in simple manufacturing technologies, especially textiles, and assembly operations has now spread all over the world. Higher level technologies, such as the ability to make internal combustion engines and simple electrical components, and to undertake fairly complex assembly operations, such as motor cars and electrical appliances, first spread to the most educated of the countries with lower labour costs, Japan, and then to its neighbours, as their skill levels improved and Japan's labour costs rose. Now countries such as Hong Kong, Taiwan, Singapore and Korea also have skilled workforces, and even Japan is surrendering its supremacy in these higher level technologies to the newcomers. The skill levels necessary to exploit other people's technological advances are now spreading to the Philippines, Indonesia, Vietnam, and the remaining countries of South East Asia, as well as to several of the South American countries.

The new competitors' products are not simply inferior imitations either; generally they are at least as good as the originals, and often are significantly better. Technology is being transferred around the world ever more rapidly, encouraged by economics, increasing skill levels and political initiatives. No one country or company has any natural permanent leadership or supremacy in any technology. Supremacy remains with those who provide the leadership in innovation into new products and processes — the number of potential imitators may multiply, but the originator will always have an initial advantage. Even the Japanese, who have been the most successful of the imitators over the past 25 years, have recognised that in the future they must achieve technological dominance through innovation and the R&D underlying it, in order to be able to lead the hordes of imitators surrounding them.

From Mechanisation to Automation

Before focusing on the possible impact of information technology on the economic development of the UK and other industrial countries, let us return to review the place of IT in an overall historical view of economic development and industrialisation. As we have remarked earlier, machines magnified the power of men's muscles, the microprocessor magnifies the power of his brain. However, this statement is too generalised for us to appreciate the role of IT, and we need a finer understanding of the significant characteristics of the industrialisation process. The industrialisation process involves several facets: one is the development of a trading economy, so that advantage can be taken of specialisation in products, services or skills — another is the organising of the production and distribution process, so that, where appropriate, technology can be brought to bear. Throughout history, one of the major factors in furthering economic development has been the application of more and more technology.

This process (the application of technology) can be broken into two major parts — the application of a machine, and the application of some form of control to a machine. A more detailed framework for these categorisations is shown in Figure 17.5, with some examples of each. Thus, the first technologies used man (or other animals) as the source of mechanical power, typically harnessed to a wheel. Each major applications area for technology, manufacturing, transportation, communications and so on, developed in this way. The next stage in development was to harness sources of natural mechanical power, the wind and falling water. The problem with these sources of power was that they were uncontrolled — if the wind dropped, or there was a drought, then the machine stopped.

Thus the first industrial revolution really took off with the advent of controllable mechanical power — first the steam engine, then the internal combustion engine, and then the electric motor. In some ways, the latter was of the greatest significance since it enabled the distribution of mechanical power to where it was needed, and the supply of an appropriate amount. The significance of this can be gauged by imagining for a moment that all of the present electric motors in factories, homes, offices, vehicles, and so on, had to be replaced by an

Characteristic	Product Examples	Power	Control
MECHANICAL			
(i) no power	plough, cart	human/animal	human brain
(ii) uncontrolled power	sailing ship watermill windmill	wind water wind	human brain
(iii) controlled power	steam train car	steam engine internal combustion engine	human brain
(iv) semi-automatic control	packing equipment assembly line	electric motor	fixed program (mechanical)/ human brain
AUTOMATION			
(v) individual automatic control	clock steam pump Jacquard loom NC machine tool robot	clockwork steam engine steam engine electric motor electric motor	fixed program (mechanical) re-programmable
(vi) full automation	automated warehouse automated factory	electric motor electric motor	integrated adaptive (feedback), re-programmable

Figure 17.5 *Mechanisation and Automation*

alternative source of power, such as a steam engine or internal combustion engine.

The next significant step was the application of some degree of automatic control. Some form of semi-automatic control, with a significant degree of human control is typified by the assembly line. Fully automatic control of a single machine (with no human control, other than turning on the power supply) dates from before the industrial revolution, in the form of the clock. But until the past 20 years, fully automatic control was confined to individual machines, and complete systems control required some significant degree of human intervention.

However, we are now entering upon the era of true automation — a word first coined in the 1930s by Ford engineers, who could conceive of a fully automatically controlled system (even though they then didn't have the means to achieve it totally), growing out of the transfer lines that they had recently developed. Automation implies that one (or more) of three additional features are present, namely:

> *a systems approach* — treating an entire process as a whole, rather than mechanising individual tasks;

> *programmability* — the ability for the control instructions in the machines (i.e. software) to be changed, so that a single machine can perform multiple tasks;

> *sensory ability* — the machine is aware of the environment within which it is operating, its relationship to other machines, and the process it is currently engaged on, so that some form of feedback or learning ability can be implemented.

All of these three features require a high degree of intelligence to be built into whatever is controlling the machines, a degree of intelligence that was generally available only with the electronic computer. However, as we have discussed earlier, at first computers were expensive, unreliable beasts, very unsuited in temperament to this area of application, and in any case, our software skills were inadequate at the time. It is only in the past decade that the hardware has become cheap enough (the concept of the computer as a component), and our software skills great enough for us to tackle this area. The cheapness and flexibility of the microprocessor is akin to that of the electric motor, delivering the intelligence required at the place where it is needed. It is the availability of cheap microelectronics which will enable true automation ultimately to come about both in the factory and the office. And again it is worth emphasising that the constraint upon this development will be the availability of people competent to use and take advantage of the new technologies.

Let us now return to consider how the economies of different countries are adapting themselves to these changes, and in particular, how we see the future development of the UK economy.

IT, Economic Development and Government Policy

In our previous discussion we examined the importance of innovation in fuelling economic development, and of R&D in driving innovation. In Figure 17.6 we detail some of the significant measures of R&D spending (which is itself notoriously difficult to measure) for the UK and some other of the major industrialised countries, for 1967 and 1975. The first columns show R&D spending as a percentage of the country's GDP — for total R&D activity and broken down into the part undertaken to meet defence needs, and the remainder. This illustrates the continuing importance of defence-related R&D spending in both the UK and USA. But in America much of this does result in development spin-offs in the civil sector and is thus not unproductive; in the UK, however, there has been almost no such spin-off. The non-defence R&D spending also illustrates how Germany and Japan have overtaken other industrialised countries in their civil R&D efforts. This is further emphasised by the final columns in Figure 17.6, which show the *per capita* industry-financed R&D, relative to the USA as a base. The UK is the only country to show a fall in such activity amongst our major competitors.

In general, this R&D spending has been oriented towards fundamental research — except in the case of Japan, where up to now it has been more towards the development of other people's fundamental research. This latter point is illustrated by the fact that the USA earns roughly 10 times more from selling technology (in the form of licences, royalties, etc), than it spends on buying it. For Japan the reverse is true — it spends about 7 times more on buying technology than it earns in selling it. For the European countries the ratio of sales to purchases of technology lies in the range from 1.0 to 1.5.

	←——R&D Spending as Percentage of GDP ——→						Per Capita Industry-Financed R&D (USA = 100)	
	Total	Defence	Other	Total	Defence	Other	1967	1975
	←——— 1967 ———→			←——— 1975 ———→				
USA	2.90	1.10	1.80	2.30	0.64	1.66	100	100
UK	2.30	0.61	1.69	2.10	0.62	1.48	78	64
Japan	1.30	0.02	1.28	1.70	0.01	1.69	48	73
France	2.20	0.55	1.65	1.80	0.35	1.45	48	55
Germany	1.70	0.21	1.49	2.10	0.14	1.96	79	94

Figure 17.6 *Spending on Research and Development*

Earlier in this chapter we suggested that the process of innovation is essentially a personal one, and is particularly one suited to the entrepreneur. Yet, over the past decade the country which has innovated most successfully has been Japan, in an environment in which both the entrepreneurial role and innovatory drive have been, to a significant extent, institutionalised into the corporate organisation. There is ample evidence that Japan intends to continue to be successful in such corporate innovation, so we consider it appropriate to review the Japanese experience.

The Japanese Experience

The major driving force in directing Japanese R&D, MITI (the Ministry of International Trade and Industry), has already indicated that it plans that by 1990 Japan's R&D spending will be equivalent to about 3% of GDP. MITI also plans to change the ratio of sales to purchases of technology — in its 'Vision of Industry in the 1980s', published in March 1980, it said that Japanese research effort should be directed more towards fundamental research — it believes that Japan is now entering an era of 'knowledge intensification'. Already there are major projects under way in computer-based pattern recognition (including voice recognition), and in completely automated batch assembly systems. Recently, the project to develop a fifth-generation computer system, drawing amongst other sources from the ideas and concepts in artificial intelligence, has been launched, with completion planned by the end of the decade.

These plans are for the future — we can perhaps gauge their likelihood of success in achieving these goals by examining past performance. As we have already illustrated in Figure 17.6, Japan's total R&D spending is still below that of other major industrial countries, expressed as a percentage of GDP. But, as we also have mentioned, this was totally oriented towards civil application. In the USA the driving force behind developments in microelectronics has been (and continues to be) the demands of the defence and space industries. The basic R&D effort in microelectronics has seen application in these areas as its initial market and paymaster — and this has generally been true in Europe too.

With the development costs written off against defence

budgets, new microelectronic products and production processes became available for commercial exploitation. In the USA the transfer of the technology from defence to civilian purposes was rapid: although some chips were application-dependent many were not, and the production technology required to make them wasn't. This was fuelled by the economics of chip production (i.e. seeking larger and larger markets); by the sheer size of the American market for new products (i.e. they have a bigger back-yard than we do); and by the great opportunity afforded by this technology to entrepreneurs in the late 1960s and early 1970s. In Europe, with very few exceptions, this technology transfer from military to civilian application did not happen, perhaps partly because of the smaller scale of the local market, partly because of the perceived poor climate for entrepreneurial risk takers.

In Japan all of the R&D in microelectronics was oriented towards the business and consumer market from the outset. The major Japanese electronics companies were taking the same technological advances in chip design and production processes from the American semiconductor companies, but developing them specifically with consumer applications in mind. Elsewhere in the world there are now perhaps only two other companies with significant consumer-electronic R&D activities — RCA in America and Philips in Holland. The result, as is by now well known, is that Japanese manufacturers dominate the worldwide consumer electronics business. What is perhaps less well known is that they are rapidly achieving the same dominance in business electronics too, becoming the purveyors of electronic *bric-à-brac* to the world.

Much of the development in Japanese consumer electronics can be traced back to programmes funded by MITI in the mid-1970s specifically oriented first to achieving parity, and secondly to overtaking the American design and production leadership in chip technology. All the major Japanese electronics companies participated in this exercise, so that all now have access to a high level of basic technological knowhow. The success of this program can be seen in the market for RAM memory chips. As we saw in Chapter 6, the Japanese suppliers gained their first major foothold in the market for memory chips with the 16-bit chip in 1979, partly by a set of fortuitous market circumstances, but also because of higher quality. As a customer of the chip makers, Hewlett Packard tested all of its incoming chips, and

Country	Supplier	Incoming Inspection Failure Rate	Field Failure (Percentage per 1000 hrs)	Quality Composite Index*
Japan	J1	0.00%	0.010%	89.9
	J2	0.00%	0.019%	87.2
	J3	0.00%	0.012%	87.2
USA	A1	0.19%	0.090%	86.1
	A2	0.11%	0.059%	63.3
	A3	0.19%	0.267%	48.1

Figure 17.7 *Comparison of Major Suppliers of 16K-bit RAMs to HP*

Note: * The Quality Composite Index is a measure of merit based on the two
failure rates, plus eight other parameters (cost, delivery, etc.) all equally
weighted

published the test results (shown in Figure 17.7) for 16K chips. This helped significantly in establishing the Japanese as suppliers of high-quality products.

But the first product resulting from the MITI project was the 64K-bit RAM memory chip, which the Japanese first began supplying in quantity in 1981, and in which they rapidly established a clear lead over the American suppliers. In the 16K chip market the Japanese suppliers have between 30 and 40% of the market, and in the 64K chip market they had over 70% of the market in 1981. The leading Japanese supplier, Hitachi, is producing about 700,000 64K-bit chips a month, more than all of the American suppliers put together. Market prices for the chip are already under severe pressure from a multitude of other Japanese suppliers entering the market, and prices may well be down to under $5 by the end of 1982 (compared with a present price of about $2 for 16K chips). Perhaps even more worrying to the American manufacturers is that the Japanese are already prototyping their 256K-bit chips.

However, when we move to consider the microprocessor market the position is different. Producing memory chips requires excellence in production technology — processor chips also require design and software skills. At present the Japanese still lag in this area and design leadership in processors still remains with the American suppliers. However, MITI has active plans in this area as a part of the fifth-generation computer system project — if they succeed and make one then they certainly have the production knowhow to make millions.

Though they might not as yet be very successful in producing general-purpose microprocessors, the Japanese consumer electronics companies have been very successful in two areas: first, in applying generally available technology to their products; and, second, in developing custom logic for their products. It is this latter capability and the superior product characteristics that are derived from it that have been a critical factor in achieving market dominance. But this has not come cheaply — in 1981 it is estimated that Sony spent about £7m. on semiconductor R&D alone, equivalent to about half of the value of its 1981 in-house chip production.

Some of the consequences of this Japanese success are illustrated in Figure 17.8, showing for manufacturing industry the changes in productivity, output and employment from 1975 to 1980 for Japan and other industrialised countries. The typical pattern has been a considerable rise in productivity, a smaller rise in output, and a fall in employment — with the UK performing worse than everyone else and the Japanese much better. The fall in manufacturing employment has led to an increase in total unemployment and a switch into service employment, and this

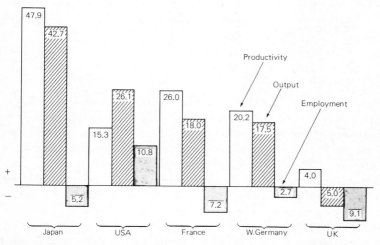

Figure 17.8 *Percentage Change in Productivity, Output and Employment, Manufacturing Industry, 1975—80*

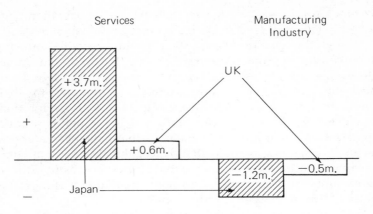

Figure 17.9 *Changes in Numbers Employed, 1973—78*

at a time when the total labour force is rising in most countries. But, if we examine the two extreme examples — the UK and Japan — in slightly more detail, a much more significant conclusion emerges. In Figure 17.9 we show the changes in numbers employed in manufacturing and services from 1973 to 1978. Although the reduction in number employed in manufacturing in Japan had been more than double that in the UK, this has been more than compensated for by an increase in service jobs almost three times bigger.

Government Policy

This brings us to one of the major conclusions of this book. In the UK, the government (of either political party) has spent enormous sums of money preserving jobs in manufacturing industry, typically in the most rapidly declining bits of manufacturing industry. In Japan the government has in total spent less, but this is directed towards encouraging innovation into new products and processes; supporting (and protecting) new industries; and developing the growth industries in the manufacturing sector. As a consequence, we have seen that manufacturing employment fell sharply, but manufacturing output rose even more sharply. The wealth that this generated was spent increasingly on services, most of which are naturally labour-intensive, employing many more people than those displaced. There is evidence that a similar demand for more services exists in the UK too only we haven't produced the goods to pay for them.

Our earlier analysis of the structure of industrial and commercial development suggested that we are about to enter into an era of widespread automation, especially in the manufacturing and assembly process; in the processing of transactions of all types; and in the functionality of most machines, whether used in the factory, office or home. We see this as one of the major consequences of the development of information technology.

This will enable new wealth to be created by employing and producing new and better products and processes; and, as has been true throughout history, this increase in aggregate wealth will be spent on improving the quality of life in all its aspects, by consuming more and different goods and services. As we said earlier, there is no indication that man's wants are satiated. And it will increasingly be the provision of services that will employ people — the machines will make the goods. But this will only happen if we create the wealth in the first place, if we adopt new techniques and practices, and make better products and processes. The evidence of the performance of the UK economy through the 1970s, and the present activities of our competitors would suggest that this innovatory process cannot be left to pure market forces alone.

The risk taken by the present attitude in the UK, by government, industry and the unions, is that we will be overtaken by intense technologically-advanced competition from abroad, resulting in loss of markets, unemployment and lower real income than we could otherwise expect. Recovery from such an occurrence will be extremely difficult in that it will require further technical innovation and change on our part, over and above that already achieved and implemented by our competitors. It would appear to us that the risks are much lower to the UK economy if we pursue a policy of actively encouraging technical innovation and change, such that UK companies and their employees can reap the benefits of the increased value-added and worldwide markets. The risks in terms of transitory unemployment are real, but would be much smaller than those faced by the present policies. Amelioration of the consequences of these risks should be the proper focus of social policy.

The review presented earlier has illustrated how Japan has

already sucessfully implemented a national policy of innovation in information technology, and has continuing plans to do so throughout the 1980s. The USA is one country relying on market forces, but the situation is not comparable to the UK. First, the sheer scale of the defence oriented R&D effort ensures such a wide spread of activity that some initiatives are bound to bear fruit (though admittedly at a very high cost). Second, the whole economic environment in the USA encourages, supports and rewards entrepreneurial activity in a way not found anywhere else in the world, and certainly not in the UK. Even in America there are now some voices expressing concern at the lead that the Japanese are achieving in chip production technology, and that they may well achieve in chip design, and suggesting that some national policy be formulated.

For example, a group of the most important American electronics companies including IBM, HP, DEC, Intel, have recently formed a joint organisation to support fundamental research in American universities on a considerable scale. The fruits of this research effort will be available to all of the participating firms. Thus the policy of other industrialised countries is that some form of coherent government plan and support in this area is inescapable — the area is simply too important to go by default. The form of the support varies; two common models are government-funded R&D, or government-created closed markets. For example, the French government as a matter of policy decided that as a part of the national plan in *informatique* it would support research in the major French electronic companies, and would use agencies such as the PTT as instruments of policy. Thus the French experiments in viewdata, electronic telephone directories, electronic mail, and so on, have resulted in large development and production contracts to French electronics companies. Some of these products are now being sold in large quantity in the American market. In both France and Germany the development and purchasing plans of the PTTs are seen as a major policy instrument, and of critical importance. Both are currently investing more than twice as much as the British Telecom in new capacity and services.

To date UK government policy has been haphazard, inconsistent and incoherent, oscillating from direct support (INMOS and the DOI support program MAPCON), to minimal support or policy (education in general), to essentially public-relations exercises (much of Information Technology Year). At one level the Department of Industry has indulged in some direct support — programmes such as MAPCON and its derivatives have helped to improve the awareness level, particularly amongst design engineers; and INMOS and (the now defunct) NEXOS were attempts to initiate a direct presence in the industry. But, especially in the context of the latter two companies, the purpose of the initiative (other than political chest thumping) does not appear to be very clear, or to have been thought through in terms of any longer term strategy. And, most significant of all, the amounts of money put into this area in total over the past five years are derisory when compared to the amounts poured into BL, British Steel and the like.

As illustrated in Figure 17.6, the UK investment in R&D does not appear particularly deficient in total when compared with other industrialised countries, but its global effectiveness does. The proportion of UK R&D going into defence-related projects is high but unlike the situation in America there has been almost no spin-off at all from this into consumer products in the UK. At present the only sign of benefit is in the satellite-building business, which is as much to do with the dynamism of British Aerospace. Generally speaking, the defence R&D in the UK has been a waste of valuable resources on an essentially non-productive activity. As the Japanese electronics companies have illustrated, ultimately R&D has to produce goods and services which people want to buy. Perhaps the most dismal comment that can be made upon the UK attitude to developments in both the UK and worldwide consumer electronics markets, is to note that the largest UK electronics company, GEC, could sit on about £600m. of liquid assets in recent years as being in its best long-term interest (as measured by short-term interest rates) — one doesn't expect to see the likes of Hitachi or Philips sitting on piles of cash!

What appears to be lacking is a clear global appreciation of the potentially critical nature of information technology — that it is not some particular fad which some short-term expediency-driven policy will hopefully submerge into obscurity. In part it is symptomatic of the general reaction and appreciation of all matters technological in the UK establishment — a great desire not to know. We consider that this is one area where the government has to lead; there has to be a review and formulation of a national policy at the highest level; only in this way can its importance and significance be signalled. The critical policy areas are threefold:

> first, a recognition of the place of innovation in information technology within the wider context of technical change; the formulation of agreed objectives between government and industry, with commensurate purpose, motivation and funding, given that the activities of our competitors mean that reliance on market mechanisms alone will not suffice; that the risks in attempting technological innovation are lower than those of doing nothing.

> second, an awareness of the long-term importance of education — to provide a climate more favourable to an acceptance and understanding of change — and the recognition that education must be a life-long experience, not a relatively short period of formal instruction.

> third, policies to cope with the social consequences; particularly the role of the government as the protector and guarantor of the rights of the individual and of the quality of society; and the need to cope with the problems of transitional unemployment arising from change, not by protecting present jobs, but by hastening the creation of new ones, and by recognising the specific problems involved.

Conclusions

The conclusions of this book can be summarised as follows:

1. Information technology is pervasive, it is *universal*; potentially it can be applied by all industrialised countries having a sufficient educational and skill level; its potential impact and importance on our future lives is too important for policy to be determined by default.

2. To the extent that the application of information technology affects relative costs, then relative prices in world trade will also be affected; the pervasive nature of information technology means that technical change in general must be the policy objective; change originating from the application of information technology is not separable or different.

3. The direct effects of technical change on domestic output and employment will be such that real income will eventually adjust, through changes in factor prices (including real wages) and industrial structure, such that labour transfers from technically obsolete organisations to those providing more goods and/or services.

4. There will be a period of transitory unemployment as a part of this adjustment process, which does not necessarily lead to cumulative unemployment, but may take a very long time to reach a new equilibrium in times of recession.

5. However, the indirect effects on output and employment could be much worse than the direct effects; serious unemployment and reductions in output (and hence aggregate wealth) may occur in an industrialised economy if an industry is subjected to intense technologically-based competition, manifest as superior products or services, or lower prices, or both.

6. Interference in the normal market adjustment processes, by propping up failing organisations, artificially maintaining employment and factor costs, is likely ultimately to prolong unemployment and heighten the risk from external technically advanced competition.

7. There is no particular distinction between the innovation of a technically advanced product, or an advanced process — both tend to occur in parallel, and both generate higher real factor incomes, demand and output.

8. The great risk exposure for the UK economy is the likelihood of being caught unprepared by major technological advances in other industrialised countries, and thus really concerns future real incomes; this risk would appear to be much higher than those arising from an attempt to generate technological leadership in domestic industries.

9. The constraint in all industrialised countries would appear to be software in the widest sense. The human capital involvement, directly in the form of knowledge and skills, and indirectly in the form of attitudes to change, will be critical and will take the longest time to create; perhaps it is here that Japan's true advantage lies.

This is the crux of this book. Hopefully you now have an understanding and appreciation of information technology and the immense but pervasive impact that it will have on our personal, corporate and national lives. In our view, information technology provides one of those few critical areas which, as a nation, we have to get right; otherwise our economic and social future is bleak.

Keeping Up-to-Date

We recognise that in the evolution of Information Technology there will be significant developments in the products available and the applications to which they are put. Therefore we list below various sources of information which we believe will help to make readers aware of these developments.

1. General Business Publications

A number of publications contain regular relevant news-items, including:

Business Week (Information Processing section)
The Economist
The Financial Times (Technology and Management pages; special surveys)
The Guardian (Futures page)

2. Management Journals

Academic management journals occasionally contain articles of interest; those listed below contain relevant material on a more frequent basis:

Harvard Business Review
Management Today
MIS Quarterly

3. Scientific Publications

Publications aimed at the lay scientist contain many articles both about the underlying technologies and specific applications; the best example is:

> *New Scientist*

4. Computer Trade Press

As in any large industry, there is a plethora of publications which chart its happenings. The majority of developments originate in America so that attention to American publications will capture these. Of the many available the most comprehensive is:

> *Computerworld*

In the UK a corresponding review is provided by:

> *Computing*

Certain publications have a different perspective, looking at specific applications, equipment and systems in greater depth; probably best known is:

> *Which Computer*

5. Personal Computer Magazines

Due to the phenomenal growth and interest in personal computing, there is now a wide range of popular magazines covering this area. The two with the best track record are:

> *Byte*
> *Personal Computer World*

6. Newsletters

In America newsletters are becoming one of the major methods by which people are being kept up-to-date in this area. Two that we have found to be particularly useful are:

> *Advanced Office Concepts*
> (published by   Advanced Office Concepts Corp.
>   One Bala Cynwyd Plaza, Suite 433,
>   Bala Cynwyd, PA 19004, USA

Microprocessors at Work
(published by   Elsevier International Bulletins,
                  Mayfield House, 256 Banbury Road,
                  Oxford OX2 7DH, England.

7. Finally, in a class of its own is a monthly American magazine which provides industry-wide application and technology reviews for the technically-literate manager, namely:

Datamation

Glossary

Access Time: the time taken for a secondary storage device to respond to a request for data by the cpu.

Acronym: a word formed from the initial letter or letters of the words in a name or phrase, e.g. Algol from ALGOrithmic Language, Cobol from COmmon Business Orientated Language.

Activity Rate: the proportion of records in a file accessed in a given period.

Actuators: output device that electronically controls a mechanical unit (e.g. water valve).

A/D: Analogue to digital conversions.

Ada: a new structured programming language being designed for the efficient development of real time systems (named after Ada, Lady Lovelace who first formulated the principles of programming).

Address: a number or label used to identify a particular computer resource, e.g. a location in primary or secondary memory.

Algol: a high-level programming language favoured by the European academic community for its structure and formality (ALGOrithmic Language).

Alphanumeric: a contraction of 'alphabetic' and 'numeric'; applies to any coding system that provides for letters, numbers, and special symbols such as punctuation.

Analogue: a signal that exactly copies another signal but in a different medium (e.g. sound by electrical voltage). Analogue signals can vary continuously in frequency and amplitude.

Analogue Computer: a computer that operates on data represented in analogue form (i.e. varying continuously). Such a computer will have a specific purpose and not be generally programmable.

Analogue Transmission: communication technique whereby signals are sent in an analogue manner, e.g. radio and TV broadcast signals.

Analyst: a person skilled in the definition of problems and the development of systems for their solution, especially systems which may be implemented on a computer; *see also* Systems Analyst.

Antiope: French Teletext System.

APL: a programming language; a high-level programming language, particularly suited to scientific applications.

Application Language: a programming language designed for a particular application area; often used directly by user, e.g. financial modelling, project planning.

Application Package: a program or set of programs designed for a particular application that can be customised to a limited extent.

Application Software: those programs designed to solve specific user problems and needs (as opposed to system software).

Architecture: *see* System Architecture.

Archive: system for storing (and retrieving) data over a long period of time. Data are usually stored off-line requiring manual intervention for retrieval.

Arithmetic Unit: the part of the central processor that performs arithmetic and logical operations.

ASCII: American Standard Code for Information Interchange. A system for coding individual characters of information using 7 bits. Used very widely on mini- and microcomputers.

Assembler: a symbolic programming language, each one of whose instructions corresponds to one of the instructions in the machine language of a computer.

ATE: *see* Automatic Test Equipment.

ATM: *see* Automatic Teller Machines.

Audit Trail: a means (such as a trail of documents, batch and processing references, log file) for identifying the actions taken in processing input data or in preparing output.

Automatic Teller Machines (ATM): electronic banking terminals for direct use by customers (including cash dispensers).

Automatic Test Equipment (ATE): equipment (usually microprocessor-controlled) capable of automatically diagnosing faults in electronic assemblies.

Automation: consideration and design systems in holistic terms. Does not necessarily imply lack of human intervention.

Back-up: alternative procedures, equipment, or systems used in case of destruction or failure of the original.

Bandwidth: the range, or width, of the frequencies available for transmission on a given channel.

Bar-code reader: a device used to read a bar code by means of reflected light, such as a scanner that reads the universal product code on supermarket products.

Baseband: data communications protocol with relatively limited band width (6 million bps). Unsuited to voice or video signals (cf. Broadband).

Basic: Beginners' All-Purpose Symbolic Instruction Code; a programming language, commonly used for interactive problem solving by users who may not be professional programmers. Very common with microcomputers.

Batch Processing: technique in data processing whereby transactions are grouped into batches before processing. Implies significant turnround time and often centralised processing.

Baud: a unit of signalling speed, in digital communications equivalent to one bit per second (note 300 baud = 30 cps).

BCD: *see* Binary Coded Decimal.

Binary: number system with a radix of 2, involving situations where there is a choice of only two possibilities. Can be represented electronically by means of a switch and thus is the fundamental basis for the stored program computer.

Binary coded: encoding data in binary/digital form for processing, storage and transmission.

Binary Coded Decimal (BCD): a coding system for numeric data.

Bit: a binary digit; (i.e. zero or one); the smallest unit of information that can be represented in binary notation.

Bits per Inch (BPI): measure of the recording density of data on a magnetic media (e.g. magnetic tape 1600 bpi).

Bits per Second (BPS): a unit used to measure the speed of transmission in a telecommunications channel.

Bootstrap Loader: a form of loader (program) whose first few instructions are sufficient to bring the rest of itself into the computer's storage from an input device.

BPI: *see* Bits per Inch.

BPS: *see* Bits per Second.

Broadband: communications protocol with high band width allowing a number of seperate communications bands to coexist, e.g. data, voice, video, etc. (cf. Baseband).

Bubble memory: memory device in which data is represented by magnetised spots (magnetic domains) that rest on a thin film of semiconductor material.

Buffer: a storage device used to compensate for a difference in rate of flow of data or in time of occurrence of events when transmitting data from one device to another.

Bundling: opposite of unbundling; the practice whereby the cost of a computer system may include not only the cpu but also an operating system, peripheral devices, maintenance agreements, etc; has largely been dropped as a result of litigation.

Bus: a passive interconnection system where the devices are connected in parallel, sharing each wire with all other devices on the 'bus'.

Byte: an eight-bit binary number which is used to represent a symbol or character (e.g. 0–9, A–Z, a–z, and special characters) for storage or transmission purposes.

Cache memory: special high speed buffer memory between cpu and primary memory.

CAD/CAM: *see* Computer Aided Design/Computer Aided Manufacturing.

CADS: *see* Content Addressable File Store.

Card Punch: a device that punches holes in cards.

Card reader: a device that translates the holes in punched cards into electrical impulses for input into the computer.

Cathode ray tube (CRT): a visual display device that receives electrical impulses and translates them into the computer.

CCITT: Consultative Committee on International Telephone and Telegraph.

Ceefax: BBC's Teletext system.

Central processing unit (cpu): the 'brain' of the computer; composed of three sections — primary memory unit, arithmetic/logic unit (alu) and control unit.

Centralised: a design alternative whereby computer power is established within a group; includes a common central database that permits authorised users to gain access to information.

Chain: a method of organising record sequence within a file by pointers indicating next and previous records in sequence.

Channel: exclusive high speed communication link between the cpu and peripheral devices.

Character: one of the symbols 0–9, A–Z, a–z, punctuation marks etc, represented usually by an 8-bit binary number. The most common coding schemes are ASCII and EBCDIC.

Characters per second (CPS): a measure of the rate of transfer of data.

Chip: a small piece of semiconducting material (usually silicon), on which are manufactured many thousands of circuits.

Circuit Switching: technique whereby the transmission path through a communications network can be determined by the switching nodes.

Clock rate: the rate of the internal synchronising clock of a processor determines the speed of processing.

CNC: *see* Computer Numeric Control.

Co-axial Cable: a cable with central core for the signal and a screening sheath around it, providing very high communications capacity.

Cobol: a high-level programming language developed for universal use in business applications (COmmon Business Orientated Language)

Collator: a mechanical device that collects together sets of information (cards/paper) in some predetermined sequence.

Common Carrier: organisation providing publicly available telecommunications links, usually the national PTT. Alternatively a communications protocol providing a common facility for sending messages.

Compatibility of Hardware: the compatibility between one generation of a manufacturer's computer system and its successor. Also the ability to interconnect equipment from different suppliers (*see* Plug-compatible).

Compatibility of Software: the degree to which programs can be run on different computers without modification.

Compiler: a translator program for high-level languages such as Fortran or Cobol; translates source program statements into machine executable or object program/code.

Components: the basic element of all electronic devices, either passive, such as resistors, capacitors etc, or active, such as transistors, valves etc.

Computer Aided Design/Computer Aided Manufacturing (CAD/CAM): integrated system co-ordinating design and manufacturing.

Computer Numeric Control (CNC): direct computer control of numerical controlled equipment (*see also* Numerical Control).

Computer Output Microfilm (COM): miniature photographic images of output. Computer output is placed on magnetic tape which serves as the input to a microfilm processor.

Concentrator: a device that systematically allocates communication channels among several terminals.

Content Addressable File Store (CAFS): an intelligent file storage device.

Control Unit: part of the processor responsible for controlling sequences of operations.

Controller: the control unit for a particular peripheral device which may contain considerable processing power and memory.

Conversational Mode: a mode of operation that implies a 'dialogue' between a computer and its user.

Core Storage: a form of high-speed storage that uses magnetic cores (now largely replaced by semi-conductor memory).

CPS: *see* Characters Per Second.

CPU: *see* Central Processing Unit.

CP/M: a widely used operating system of microcomputers (originally 8 bit but now 16 bit) which specifies a standard for memory layout, and file storage.

Crash: general term indicating any breakdown or malfunction of a computer system.

CRT: *see* Cathode Ray Tube.

Custom Logic: *see* Digital Logic.

Cycle Time: time taken to fetch and return an item from primary memory.

Cylinder: a vertical grouping of each track in the same relative position on each platter of a multi-platter magnetic disk.

Daisy Wheel Printer: slow speed (less than 100 cps) high quality printer where the characters are positioned on a wheel at the end of a circle of protruding stalks (i.e. daisy like).

DASD: *see* Direct Access Storage Devices.

D/A: Digital to analogue conversions.

Data: basic, raw, unprocessed information. Usually bearing direct relationship to facts, events or activities. Data has to be encoded before it can be processed by a computer.

Data Dictionary: Part of DBMS that contains all the definitions of the data items and their relationships.

Data Havens: countries with a very liberal attitude to data protection.

Data Items: smallest useful unit of information.

Data Processing (DP): the organisation of the processing of business records and information; a generic term for computing in organisations.

Data Protection: concerned with the contents and accessibility of computerised information.

Data Protection Agency (DPA): a quasi-government agency set up to administer data protection legislation.

Data Set: *see* File.

Databank: a structured collection of information usually referring to a specific topic. Fairly static in nature with information being steadily added.

Database: collection of related files usually of a volatile nature.

Database Management System (DBMS): part of the system software that manages the interface between the filing system and the application program. Also containing a set of utility programs to access the data directly (*see* Query Language, and Report General).

DBMS: *see* Database Management Systems.

DDP: *see* Distributed Data Processing.

Debug: to isolate and remove all errors from a program.

Digital: representation of data by a numeric value.

Digital Computer: general purpose computer using binary logic (cf. analogue computer).

Digital Logic: special purpose circuits that act upon information in binary coded form.

Digital Transmission: communications technique whereby signals are transmitted in binary form. Thus analogue signals must first be digitalised (A/D) before transmission and then reconverted (D/A).

Direct Access Processing: method of file processing where transactions can be processed in any sequence.

Direct Access Storage Device (DASD): a storage device on which information can be stored and retrieved in a direct, non-sequential manner, such as a disk drive.

Disk Drive: a mechanical device used to record and retrieve data from a magnetic disk. Includes a rotational unit and read/write head.

Disk Operating System (DOS): an operating system including the ability to handle disk drives; a relatively primitive system; *see also* Operating System.

Disk Pack: a stack of magnetic disks.

Diskette: *see* Floppy Disk.

Distributed: a design alternative whereby each business unit has its own computer power although total organisation-wide control exists.

Distributed Data Processing (DDP): organisation of data processing resources in which some capabilities are brought closer to the place where the data originates and/or processing is required.

Distributed System: equipment configuration characterised by geographically dispersed computers linked together in a communication network; user-orientated but under total organisation-wide control.

Documentation: collection of written information describing the internal systems and external (user) operations of a program.

DOS: *see* Disk Operating System.

Downtime: the elapsed time when a computer is not operating correctly due to machine failure.

DP: *see* Data Processing.

DPA: *see* Data Protection Agency.

Driver: part of operating system software responsible for controlling the basic operations of an individual device.

EBCDIC: Extended Binary Coded Decimal Interchange Code. IBM standard coding system now used widely by other manufacturers for exchanging data between different computers.

EDP: electronic data processing: *see* Data Processing.

EFT: *see* Electronic Funds Transfer and Electronic Financial Transactions.

Electronic Financial Transactions: initiation of a financial transaction by electronic means; does not necessarily include electronic funds transfer.

Electronic Funds Transfer: transfer of funds by electronic message switching systems; *see also* SWIFT.

Electronic Mail: facility to send documents in electronic form ranging from the ability to interconnect word processors (*see* Teletex) to networks where documents/messages may be stored and subsequently forwarded or retrieved.

Electrostatic Printer: a non-impact printer in which electromagnetic impulses and heat are used to affix characters to paper.

Emulator: a program that instructs a computer to act as if it were another model of computer thus enabling object programs to operate on a different computer.

Encryption: coding or scrambling of information to prevent unauthorised use.

EPROM: Erasable Programmable Read Only Memory; *see* PROM and ROM.

Etching: stage in manufacturing process of integrated circuits where acid is used to remove unwanted layers.

Ethernet: local area network developed by Xerox using a baseband approach.

Executive: *see* Operating System.

Expert System: system for encoding knowledge in certain specialised domains capable of making inferences deduced from information and of having an English-like dialogue with the user.

Fascimile (FAX): technique of sending a pictorial copy of a document over a telecommunications link.

FET: Field Effect Transistor; one type of microelectronic device.

Field: a meaningful item of data, such as a social security number.

File: a grouping of related records, sometimes referred to as a data set.

Flexible Manufacturing System (FMS): an integrated manufacturing system (usually computer controlled and including robots) capable of producing small batches of goods.

Floating Point: method of storing numbers which contain decimal parts and have a wide range. The number is stored as a mantissa (fixed point part) and exponent (the number of decimal places).

Floppy Disk: also called a diskette or flexible disk, a low-cost random access form of data storage made of vinyl and at present having a storage capacity of up to 1 Mbyte.

Flowchart: the program flowchart which is a graphic representation of the types and sequences of operations in a program, and the system flowchart which shows the flow of data through an entire system.

FMS: *see* Flexible Manufacturing Systems.

Fortran: a programming language whose name is the abbreviation of FORmulation TRANslation, particularly suitable for the processing of scientific applications.

Frame: a screen-full of information for storage and display in a videotext or teletext system.

Gate Array: general purpose digital logic which is customised only at the final stage of manufacture. Also called uncommitted logic array.

Gateway: access point in some network to other networks or systems specifically referring to Prestel as a method of accessing third party computers.

Generation: term describing stages in development of computer systems. State of the art computers are considered to be fourth generation. Japanese are attempting a major project to develop the fifth generation.

Hard-copy: printed output.

Hardware: the electrical circuitry and physical devices that make up a computer system.

Hashing Algorithm: the process whereby a record key is turned into a disk address in direct access file processing.

Head Crash: when the read/write head on a magnetic disk touches the platter surface data are irretrievably lost and considerable damage can be caused to the disk drive.

Hex: hexadecimal is a number system to the base 16. It is used by programmers to look at the contents of computer memory in a more convenient form than binary.

High-level Languages: English-like coding schemes which are procedure problem and user oriented. Must be translated to machine code instructions before computer can obey them.

Impact Printer: a printer that forms characters by physically striking a ribbon against paper.

Index: an ordered reference list of contents of a file, or the keys for identification or location of the contents.

Index-sequential Processing: a file processing technique in which records are placed on a file in sequence and multiple level index is maintained, thus allowing both sequential and direct access processing (ISAM).

Information Suppliers: organisation that provides information for sale to subscribers of British Telecom's Prestel service.

Information Resource Management (IRM): theory of management that identifies information as a key resource and treats it like other managed resources (i.e. finance, marketing, personnel etc.).

Information System: an integrated network of personal equipment and procedures designed to satisfy the information requirement of management.

Information System Manager: the manager responsible for planning, organising, staffing and controlling all data processing activity.

Ink-Jet Printer: a non-impact printer that uses a stream of charged ink to 'paint' characters.

Input: data that are submitted to the computer for processing.

Input/Output (I/O): a general term for the equipment used to communicate with a computer and the data involved in the communication.

Input/Output (I/O) Devices: devices capable of both transmitting input to a computer, and of receiving output back from it.

I/O: *see* Input/Output and Input/Output Devices.

Instruction: a statement that specifies an operation to be performed and the associated values or locations.

Instruction Register: a register where each instruction is decoded by the control unit.

Instruction Set: the fundamental logical and arithmetic procedures that the computer can perform, such as addition, subtraction and comparison. Specific to a given computer.

Integer: method of representing a whole number. Positive numbers are simply their binary value while negative numbers are stored in '2s complement form'.

Integrated Circuit: a small chip, usually of silicon, which contains a large number of interconnected electronic components.

Intelligent Terminal: a terminal with an internal processor that can be programmed to perform specified functions, such as data conversion, and control of other terminals.

Interactive Computing: conversation (interaction) between a user and a central computer via an on-line terminal.

Interface: a shared boundary, e.g. the boundary between two systems, or between a computer and one of its peripheral devices.

Interpreter: a system program that interprets a high-level language instruction one at a time. Slower than a compiler in computing terms but potentially much faster for system development.

Interrupt: the ability of a processor to suspend operation on one process in order to service the requirements of some higher priority requests.

IRM: *see* Information Resource Management.

Job Control Language (JCL): a high-level language used by programmers to instruct the operating system.

Josephson Junction: a semiconductor device operated at cryogenic temperatures (i.e. temperatures close to absolute zero), making use of the physical properties of individual atoms and molecules, instead of the bulk properties of the materials used. Such devices offer extreme packing densities and switching speeds well beyond current capability.

K: a symbol used to denote 1000 (strictly 1024) where referring to specific capabilities of a computer system.

Key: a unique identifier for a record; used to sort records for processing or to locate a particular record within a file.

Key to disk: hardware designed to transfer data entered via keyboard to magnetic disk or diskette.

Keyboard: input device like a typewriter usually part of visual display unit or terminal.

Keypunch: a keyboard device that punches holes in a card to represent data.

Keyword: a word or phrase in a document that has been designated as a key for retrieval. A more sophisticated approach is using keywords in context (KWIC).

LAN: *see* Local Area Network.

Laser Printer: a type of non-impact printer that uses a laser to produce images on the drum of an electrostatic printer/copier, and thus on to paper.

Laser Storage System: a secondary storage device using laser technology to encode data on to an optical disk; *see also* Optical Disk.

LCD: *see* Liquid Crystal Display.

Leased Line: a service offered by the common carriers in which a customer may lease, for his exclusive use, a circuit between two or more geographic points.

LED: light emitting diode.

Light Pen: a pen-shaped object with a photoelectric cell at its end; used to draw lines on a visual display screen.

Liquid Crystal Display (LCD): uses the property of certain materials to polarise light by applying an electric potential.

Local Area Network (LAN): a technique for providing an internal communications ring main.

Local Loop: connection between a local exchange and subscribed on the switched telephone network.

Location: generally, any place in which data may be stored.

Log: a record of the operations of data processing equipment, listing each job or run, the time it required, operator actions, and other pertinent data.

Logging: keeping a copy of transactions on a separate file for audits and security purposes.

Login: introductory procedure when accessing an on-line computer system.

Loop: a sequence of instructions that can be executed repetitively with modified sets of data. Also describes the situation when a program goes into an 'infinite loop' and must be interrupted by the user.

Low-level language: programming languages where each instruction maps on to a single machine code instruction.

LPM: lines per minute.

LST: Large Scale Integration. A measure of the packing density on a chip.

Machine Code: the only set of instructions that a computer can execute directly; a cycle that designates the proper electrical states in the computer as combinations of 0s and 1s.

Machine Readable: information stored in such a manner that it can be automatically interpreted by a computer e.g. magnetic tape, punched cards.

Macro: a facility for defining a procedure that may subsequently be invoked as a simple command.

Magnetic Core: an iron-alloy doughnut shaped ring about the size of a pin head of which memory can be composed; an individual core can store one binary digit; a technology used in second-generation computers, now overtaken by semiconductor RAM memories.

Magnetic Disk: a storage medium consisting of a metal platter coated on both sides with a magnetic recording material upon which data are stored in the form of magnetised spots (a DASD).

Magnetic Tape: a storage medium consisting of a narrow strip upon which spots of iron oxide are magnetised to represent data; a sequential storage medium.

Mainframe: type of large scale computer built on a large frame or chassis. Used by large centralised dp department.

Management Information System (MIS): a formal network that extends computer use beyond routine transaction processing and reporting, and into the integration of applications and the area of management decision making; its goal is to get the correct information to the appropriate manager at the right time.

Manual: documentation provided by the originators of a particular system or piece of software to explain methods of operation.

Mass Storage Devices: a class of secondary storage devices capable of storing very large volumes of data; offers cost advantages over disk storage, but its much slower retrieval time is measured in seconds.

Master File: a file that contains relatively permanent data; updated by records in a transaction file.

Matching: comparing master records with transaction records to see if certain fields are identical.

Mechanisation: the application of computer related equipment to improve the productivity of specific tasks.

Memory: part of the computer that stores binary-coded information; primary is working storage; secondary is permanent storage.

Merge: to form a single sequenced file by combining two or more similarly sequenced files.

Message Switching: a technique consisting of receiving a message at a connecting point in a network, storing it until the appropriate output circuit is clear, and then retransmitting it.

Microcomputer: a computer taking up a small amount of space and relatively inexpensive. Usually suited to running only single applications.

Microelectronics: the technique of manufacturing many different inter-connected electronic components in a small physical space.

Microprocessor: the arithmetic/logic unit and control unit, some limited memory, and limited I/O capability of a processor, all manufactured on a single chip.

Microsecond: one millionth of a second; term used in specifying the speed of electronic devices.

Millisecond: one thousandth of a second; a term used in specifying the speed of electronic devices.

Minicomputer: a medium scale computer suited to supporting inter-active applications for a large number of users.

MIPS: millions of Instructions Per Second; a measure of processing power. A modern large computer would be rated at over 10 MIPS.

Modem (data set): a device that modulates and demodulates digital signals transmitted over analogue communications facilities.

Modular programming: a programming approach that emphasises the organisation and coding of logical program units, usually on the basis of function.

Modulation: a technique used in modems (data set) to transform digital signals to analogue form.

Module: a part of a whole; a program segment; a sub-system.

Monitor: *see* Operating System.

MOS: metal oxide semiconductor. Technique for building micro-electronic circuits.

MSI: Medium Scale Integration. Referring to the packing density of integrated circuits.

Multiplexor: a device that permits more than one terminal to transmit data over the same communication channel.

Multiplexor Channel: a controller that can handle more than one I/O device at a time; normally controls slow-speed devices such as card readers, printers or terminals.

Multiprocessor System: a multiple cpu configuration in which jobs are processed simultaneously.

Multiprogramming: a technique whereby several programs are placed in primary storage at the same time.

Name-linked Files: data files one of whose keys is a person's name.

Nanosecond: one billionth (thousand millionth) of a second; a measure of the speed of electronic devices.

NC: *see* Numerical Control.

Network: the interconnection of a number of locations or devices by communications facilities; *see also* Public-Switched Network.

Non-impact Printer: a hard-copy output device that forms the images on paper by non-impact methods (e.g. electrostatic or ink-jet printers).

Non-procedural Software: software designed to cope with a general class of user processing requests in terms of 'what' the user wants, rather than 'how' a specific request is to be handled (e.g. Query Languages; Report Generators).

Numerical Control: control of a piece of equipment (usually a machine-tool) by means of binary-coded data, usually prepared on paper-tape.

Object Program/Code: a sequence of machine-code instructions arising as the output from a compiler program with a source program/code as input.

OCR: *see* Optical Character Recognition.

Octal: a number system to the base 8; it is sometimes used as a more convenient way of representing binary numbers; *see also* Hex.

OEM: *see* Original Equipment Manufacturer.

Off-line: equipment or devices not under the direct control of the central processor.

On-line: equipment or devices in direct communication with the central processor.

Operating System (OS): a collection of programs designed to enable a computer system to manage itself, with minimum human intervention.

Optical Character Recognition (OCR): devices capable of 'reading' characters on paper and converting the optical images into binary-coded data.

Optical Disk: a system for storing binary-coded information optically rather than magnetically, with improvements in speed and recoding density.

Optical Fibres: fibres of finely-drawn, flexible, highly-reflective glass, down which pulses of light may be transmitted representing binary-coded data.

Oracle: the Independent Broadcasting Authority's (IBA's) Teletext system.

Original Equipment Manufacturer (OEM): a firm producing computing or telecommunications hardware, typically sold through third parties (dealers, system houses) to the final customer.

OS: *see* Operating System.

Output: information that arises as a result of some processing.

Output Device: a device capable of transforming binary-coded information either into human-interpretable form (e.g. hard-copy, display or voice), or into a form to control some mechanical device; *see* Actuator.

PABX: *see* Private Automatic Branch Exchange.

Packet: a standard-sized unit (in terms of numbers of bits) into which a message is broken down for transmission using a packet switching protocol.

Packet Assembler/Dis-assembler (PAD): a device capable of breaking a message into standard 'packets' for transmission using a packing switching protocol; and of re-assembling 'packets' into a message.

Packet-Switched Network (PSN): a telecommunications network operated using packet-switching protocols; may be a separate physical network, or a virtual network on an existing real network; may be public or private system.

Packet Switching: a transmission method in which the messages containing the data are broken down into fixed-length 'packets'; each packet contains routing information and shares the network with other packets, improving the overall utilisation of the network.

Packing Density: a measure of the closeness with which circuits may be placed to each other on the surface of a chip.

PAD: *see* Packet Assembler/Dis-assembler.

Pages: (1) in a virtual-storage environment, portions of programs or data which are kept in secondary memory and are loaded into primary memory when needed for processing. (2) in a viewdata environment, a screen-full of information to be displayed as such; *see* Frame.

Paging: a method if implementing a virtual-storage system; programs and data are broken down into fixed size pages (or segments) in secondary memory, and are loaded into primary memory when needed for processing.

Paper Tape: a data recording medium in which characters are represented by horizontal patterns of holes punched across the width of the tape.

Paper Tape Reader: a device capable of sensing the pattern of holes punched in a tape, and of converting them to appropriate binary-coded signals.

Parallel: processing or transmission of instructions or data simultaneously (as opposed to serial).

Parity: error detection technique using one bit of a byte or word.

Partition: in a multiprogramming environment, the primary memory area reserved for one program; may be fixed or variable.

Pascal: a high-level programming language, particularly designed to aid the implementation of structured design.

PCM: *see* Plug-Compatible Manufacturer.

Peripherals: auxiliary computer equipment connected to the central processing unit (e.g. I/O devices, secondary memory devices).

Photolithography: a method of achieving a two-dimensional image on a surface by using an optical mask; in particular, defining the pattern for electronic circuits on the surface of a chip.

Picosecond: one thousand thousand millionth of a second; a measure of the speed of electronic devices.

Platter: the flat recording surface of a magnetic disk; the disk itself.

PL1: Programming Language One; a high-level programming language, designed by IBM for both business and scientific applications.

Plotter: an output device that produces hard-copy graphical images.

Plug-compatible: devices (usually peripherals) which are available from another company and are interchangeable with the devices from the original manufacturer.

Plug-Compatible Manufacturer (PCM): a manufacturer of hardware which is plug-compatible with equipment from the original manufacturer.

Pointer: a link used to indicate the location of another record (usually the next or previous in logical sequence) in an index-sequential file.

Point-of-Sale (POS) Terminal: a device incorporating a cash-register which can also capture the transaction details electronically and store and/or forward them.

Portability: an attribute of a program which can be transferred to another computer system and run without modification to the program or its data.

POS: *see* Point-of-Sale Terminal.

Prestel: British Telecom's public videotext service.

Primary Memory: the part of the central processing unit which holds instructions for execution, data being processed; also known as primary storage, main memory or storage; typically made up from RAM chips.

Printer: an output device producing hard-copy.

Privacy: an individual's right to knowledge and control of personal information recorded on (computerised) files.

Private Automatic Branch Exchange (PABX): the internal telephone exchange for an organisation.

Private Leased Line: *see* Leased Line.

Procedure: a set of rules for achieving some defined goal.

Processor: an electronic device capable of interpreting and executing binary-coded instructions.

Program: a set of instructions which tell a processor how to perform a specific task; of two types: application, concerned with end-user orientated tasks; and system, concerned with internal control and scheduling tasks.

Programmable: a procedure capable of being set out as a series of logical steps, in an unambiguous, deterministic manner; capable of being programmed.

Programmable Read-Only Memory (PROM): a read-only memory (ROM), capable of being re-programmed once by the user.

PROM: *see* Programmable Read-Only Memory.

Protocol: a set of agreed rules by which communications may be established between devices, providing certain defined facilities.

Prototyping: a process for system development aimed at implementing a useable first-stage system at minimum cost and time; implies feedback, learning and an interactive process.

PSN: *see* Packet-Switched Network.

PTT: Post, Telephone and Telegraph; a generic name, typically used in Europe to refer to the telecommunications authority.

Public-Switched Network: the standard public telephone network provided in the UK by British Telecom.

Punched Card: a data recording medium in which characters are represented by vertical patterns of holes punched in the card; the basis for unit-record equipment.

Punched Card Reader: a device capable of sensing the pattern of holes punched in a card, and of converting them to binary-coded signals.

Query Language: software capable of handling enquiries expressed in English-language form on a database; an example of non-procedural software.

RAM: *see* Random Access Memory.

Random Access Memory (RAM): (1) memories giving a constant access time to any location independent of the last location addressed (as opposed to sequential memories). (2) a generic name for read/write memory (as contrasted to read-only memory), typically forming the primary memory of a computer system; presently implemented using RAM chips.

Random File: a method of file organisation to give known access time to any record in the file, when processing sequence is unknown.

Random Processing: a method of record processing in which retrieval of records is independent of sequence; transactions can be processed directly; average access time is fast and consistent, but processing is relatively expensive for high activity and volatility; *see also* Sequential Processing, and Index-Sequential Processing.

Read: to copy the contents of some storage media containing binary-coded information; the storage contents are not destroyed by the reading process; *see also* Write.

Read-Only Memory (ROM): Memory device from which the data can only be read and which cannot be written to; memory whose contents are fixed at time of manufacture; presently implemented using ROM chips.

Read/Write: an attribute of a device by which data (or instructions) may be both copied, i.e. 'read', and to which data (or instructions) may be stored, i.e. 'written', overwriting the previous contents.

Read/Write Head: an electromagnet used as a component in magnetic recording devices (e.g. tape or disk), either to sense a magnetised area (i.e. read) or to magnetise an area (i.e. write).

Real: a physical device or entity (as opposed to virtual).

Real Time: a term signifying that the results of some processing step will be available in a short enough time to influence or control the procedure or task which is the subject of the processing.

Record: a set of logically-related items pertaining to a single entity, which are retrieved from or updated in a file in a single operation; the logical unit within a file.

Recording Density: the number of binary digits stored within some unit length or area; typically measured as bits-per-inch on magnetic tape or disk.

Register: a dedicated memory location within the central processing unit used as immediate storage by the arithmetic and control units.

Remote Access: access to a computer system via some telecommunications link.

Remote-Job-Entry (RJE): a method for entering jobs for processing into the normal batch-processing queue of a central facility from some remote location.

Report Generator: software capable of producing reports from a database in response to instructions from the user; an example of non-procedural software.

Reprographics: equipment concerned with the reproduction of images on hard-copy; e.g. electrostatic copiers.

Response Time: the time that elapses between entering the last item of input into a processing task and the start of the output.

Retrieval: location of a specific item of information fulfilling some given criteria in a file, and copying it into primary memory.

Ring: a form or organisation of a communications network, especially applicable to local area networks; involves devices connected to a single transmission line in a ring arrangement; organised on some token-passing protocol.

Robots: general purpose programmable device of human scale and facility, with some capability for sensing its environment.

ROM: *see* Read-Only Memory.

RPG: Report Program Generator; a business-oriented high-level programming language.

RS 232: a standard communications interface for connecting low-speed serial devices; especially used in connecting terminals to computer systems.

Run: to initiate the execution of a specific program (or the one presently resident in primary memory).

Secondary Memory: memory used for long-term, large-scale storage of files; typically comprising magnetic tape or disk; also known as secondary storage or auxiliary storage.

Second-Generation Computer: a computer using transistors as its basic component technology; faster, smaller, more powerful and more reliable than first-generation; higher-level programming languages and operating systems available.

Second Sourcing: an agreement whereby the original manufacturer of a chip design licenses other manufacturers to produce the same design and provides production masks.

Security: precautions in a computer system to guard against unauthorised access, crime, natural disasters and accidental damage.

Segment: *see* Paging.

Segmentation: *see* Paging.

Self-documenting: a facility within a piece of software for providing several levels of explanation.

Semiconductor: any material which acts as an electrical conductor when the voltage is above a certain level, and as a resistor when it is below it (i.e. it can act as an electronic switch).

Semiconductor Memory: device capable of storing binary-coded information electronically; e.g. RAM or ROM.

Sensors: devices capable of sensing and/or measuring a change in status, and of transmitting that information to some controller.

Sequential Access: a method of retrieving data in which records have to be searched from the beginning of the file for those required.

Sequential File: a method of file organisation in which records are stored in a known sequence which will be used for retrieval.

Sequential Processing: a method of record processing in which records are retrieved in a known sequence; transactions to be processed have to be sorted in the same sequence prior to processing; average access times are slow, but processing is economical for high activity rates; *see also* Random Processing, and Index-Sequential Processing.

Serial: processing or transmission of instructions or data in strict sequence, i.e. one at a time; not parallel.

Service Bureau: an organisation supplying computing services to other organisations; a computer utility.

Shared-Logic: an application implemented as a part of a general multiprogramming system; sharing resources between several applications (i.e. not stand-alone).

Silicone chip: *see* Chip.

Slave: a device whose capabilities are totally determined by some master control unit.

Small Scale Integration (SSI): a measure of the packing density of circuits of a chip; the lowest such measure.

SNA: *see* System Network Architecture.

Software: a collection of programs and routines providing instruction to a computer system; primary classification into application and system.

Software Package: a collection of programs and routines designed to provide a solution to problems in a specific area.

Sort: to arrange items of information into specific sequence (e.g. ascending alphabetic).

Sorter: a piece of unit-record equipment designed to sort records into a specific sequence.

Source Language: the language in which the problem statement is encoded (e.g. Cobol, Fortran, Basic, etc).

Source Program: a program written in a specific source language; the input to the compiler program; *see also* Object Program and Compiler.

SPC: *see* Stored Program Control.

Spooling: a technique for removing the direct connection between I/O devices and the central processor, in order to improve system throughput.

SSI: *see* Small Scale Integration.

Stand-Alone: a processing unit not connected to any other system, typically dedicated to a single application; *see also* Shared Logic.

Statement: a single instruction in a high-level language program.

Storage: a device capable of retaining information in binary-coded form for subsequent retrieval; synonymous with memory.

Store and Forward: an attribute of a message-handling system, such that it can accept and store messages to be forwarded when the addressee is active.

Stored Program Control (SPC): a processing unit operating under the control of instructions stored in its own memory, which may be modified as a result of processing by the host processor, or any other to which it can communicate.

Structured Design: a top-down modular approach to system design.

Structured Programming: a top-down design method applied to programming; imposes a formal structure on the design of programs which enables the correct functioning of the program to be established quicker.

Supervisor Program: the major component of an operating system, co-ordinating the other parts of the system and scheduling tasks; also referred to as an executive or monitor.

Swapping: in a virtual storage environment, the process of transferring a block of program or data (*see* Page) from secondary memory (virtual memory) to primary memory (real memory), and *vice versa*.

SWIFT — Society for Worldwide Interchange of Financial Transactions: an interbank organisation set up to manage and operate a worldwide electronic network for funds transfer between the participants.

Switch: a device capable of making or breaking a connection; a part of processing logic, or a telecommunications network.

Systems Analysis: the process of analysing a problem into its basic operations and specifying these in a programmable form.

Systems Analyst: one who carried out the systems analysis function.

System Architecture: a term designating the structure of a computer system or communications network.

System Design: the process of designing a processing system, with associated data capture, storage and output, to meet the need identified in the systems analysis phase within systems development.

System Development: the process of identifying a problem; evaluating alternative approaches to its solution; designing the chosen system; implementing and evaluating it.

System Network Architecture (SNA): a protocol for handling computer to computer communications; originated by IBM.

System Software: a collection of programs designed to control the computer system; major components are: controllers for the major hardware components, compilers and other utility programs, and a supervisor to manage the system and schedule tasks.

Tabulator: a piece of unit-record equipment capable of cumulating the number of occurrences of a particular set of characters in a group of records.

Tape Drive: a device which moves magnetic tape past a read/write head enabling binary-coded information to be written to or read from the tape.

Telecommunications: a system incorporating the combined use of computer and telecommunications facilities.

Télétel: French PTT videotext system.

Teletex: an internationally agreed standard for message handling between simple terminal devices and word processors over the public-switched network.

Teletext: broadcast information system utilising spare bandwidth on the television broadcast frequencies; *see also* Ceefax and Oracle.

Teletype: a trademark of the Teletype Corporation; a low-speed hard-copy terminal device.

Telidon: the Canadian PTT's videotext system.

Terminal: an input/output device connected to a computer system via a communications link; originally low-speed and hard-copy.

Text: information stored and processed as characters, rather than numbers.

Text Processing: a system designed to facilitate the processing of large volume of textual material (e.g. articles, reports, books etc).

Thrashing: the situation occurring when a virtual system is excusively swapping programs.

Throughput: a measure of the work done by a computer system over some specified time interval.

Time-sharing: a form of operating system which allows two or more users to share central resources, yet apparently receive simultaneous service.

Top down: methodology for approaching system and program design by breaking the problem down into levels, increasing detail.

Touch-tone: a device (usually a telephone with a key-pad rather than dial) capable of sending a coded sequence of tones over the public-switched network.

Track: the logical areas in which information is recorded on magnetic media; bands stretching the length of the tape, or concentric bands on disk.

Transaction: an event which generates one or more items of information which are captured and stored for processing.

Transaction File: a file containing transactions captured over a specific period to be processed against some master file(s).

Trans-border Dataflow: transmission of information in electronic form over national boundaries.

Transistor: a technological stage in the development of electronic devices; superior to valve technology, but inferior to integrated circuits (chips); the technology of second-generation computer systems.

Transmission Media: the media used to form the transmission link between two points (e.g. copper cable, fibre optics, microwave).

Transpac: the implementation of a public packet-switched network by the French PTT.

TTL: Transistor-Transistor Logic; a method of forming electronic circuits on chips.

ULA: *see* Uncommitted Logic Array.

Unbundling: the practice of charging a separate price for each individual component of a computer system, whether hardware or software; resulted from US Justice Department action against IBM, and created the opportunity for plug-compatible manufacturers.

Uncommitted Logic Array (ULA): a form of logic circuit on a chip where the final purpose of the logic can be determined at the final stage of manufacture, thus giving economies of scale in earlier stages.

Unit Record Equipment: equipment designed to process one physical record at a time; typically stored on punched cards.

Update: to incorporate into a master file the transaction data that has arisen since the last update.

Upgrade: replacing an existing computer system by a more powerful (hopefully compatible) model in the manufacturer's range.

Upward-Compatible: the ability to move to a larger, more powerful computer system without having to amend or modify software.

Utility Programs: a set of programs performing commonly required tasks (e.g. sorting, merging, editing etc); usually part of the operating system.

Valve: the original technology for manufacturing electronic components; inferior to transistor technology; characteristic technology of first-generation computer systems.

Validate: to check on the accuracy and authenticity of data capture or processing.

Value-Added Network (VAN): a logical, virtual network created on a real network (typically the public-switched network), in order to market a specific service.

VAN: see Value-Added Network.

VDU: see Visual Display Unit.

Verify: to check the accuracy of a transcribing operation, usually by a comparison procedure.

Very Large Scale Integration (VLSI): a measure of the packing density of circuits on a chip; presently the highest measure of density.

Video Disk: storage of video images on an optical storage device.

Videotext: a system specifying standards for the storage of data in content-independent form, and for accessing such systems over telecommunications networks; see Prestel, Telidon.

Viewdata: see Videotext.

Virtual: implementation by software of physical resources (e.g. storage or processor); as opposed to real.

Visual Display Unit: an input/output device on which output is displayed on a cathode-ray tube; input is by means of keyboard.

VLSI: see Very Large Scale Integration.

Voice Message System (VMS): a system for the capture, storage and subsequent onward transmission of voice messages in binary-coded form.

Volatility: a measure of the rate of addition/deletion of records in a file.

Wafer: a slice of the semiconductor material (usually silicon) on which the chips are manufactured; wafers are the basic production unit for much of the chip-production process; there are multiple chips per wafer.

Winchester: a disk-drive technology in which the platters and read/write heads are enclosed in a hermetically sealed unit at time of manufacture; originally developed by IBM.

Wired Society: concept of a society organised around tele-processing networks for communications and transaction processing.

Word: a group of bits treated as a whole by the processor; a measure of the power of a processor.

Word Length: the number of bits in a word.

Word Processing: a system specifically designed for the manipulation of text, both as individual characters and as logically related groups of characters (i.e. words and phrases).

Work Station: a device designed as a user interface to a computer system.

Working Storage: an area (in primary memory) set aside by the programmer for local use within the program to hold non-permanent data.

Write: the process of recording information in a memory location, or on a secondary memory media (i.e. tape or disk); writing destroys the previous contents of the memory location or memory area.

Xerographic Printer: a printer using methods similar to those of a xerographic copier to produce images on paper.

X.25: an international standard (from the CCITT) for packet-switched network protocol; adopted by PTTs and majority of computer manufacturers.

Yield: a measure of the proportion of good chips produced on a wafer.

Index